Quantum Chemistry

Quantum Chemistry

Dr. M. S. Prasada Rao
Professor of Chemistry

Andhra University

CRC Press is an imprint of the
Taylor & Francis Group, an **informa** business

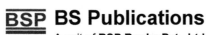

A unit of **BSP Books Pvt., Ltd.**
4-4-309/316, Giriraj Lane, Sultan Bazar,
Hyderabad - 500 095

First published 2023
by CRC Press
4 Park Square, Milton Park, Abingdon, Oxon, OX14 4RN

and by CRC Press
6000 Broken Sound Parkway NW, Suite 300, Boca Raton, FL 33487-2742

© 2023 BSP Books Pvt. Ltd

CRC Press is an imprint of Informa UK Limited

The right of M. S. Prasada Rao to be identified as author of this work has been asserted in accordance with sections 77 and 78 of the Copyright, Designs and Patents Act 1988.

All rights reserved. No part of this book may be reprinted or reproduced or utilised in any form or by any electronic, mechanical, or other means, now known or hereafter invented, including photocopying and recording, or in any information storage or retrieval system, without permission in writing from the publishers.

For permission to photocopy or use material electronically from this work, access www.copyright.com or contact the Copyright Clearance Center, Inc. (CCC), 222 Rosewood Drive, Danvers, MA 01923, 978-750-8400. For works that are not available on CCC please contact mpkbookspermissions@tandf.co.uk

Trademark notice: Product or corporate names may be trademarks or registered trademarks, and are used only for identification and explanation without intent to infringe.

Print edition not for sale in South Asia (India, Sri Lanka, Nepal, Bangladesh, Pakistan or Bhutan).

British Library Cataloguing-in-Publication Data
A catalogue record for this book is available from the British Library

Library of Congress Cataloging-in-Publication Data
A catalog record has been requested

ISBN: 9781032406374 (hbk)
ISBN: 9781032406381 (pbk)
ISBN: 9781003354048 (ebk)

DOI: 10.4324/9781003354048

Typeset in Times New Roman
by BSP Books, Hyderabad 500 095

Dedicated to
my teacher and mentor
Late Prof. S. R. Sagi
School of Chemistry, Andhra University

Preface

Chemistry is an experimental science. Nevertheless, quantum chemistry is quite different and remained an enigma to most of the chemists and students of chemistry. This is mainly because its foundation lies in 'quantum mechanics', a collection of abstract laws and equations. Therefore, there is a need to bring the students out of this frame of mind and make them look beyond and enjoy the beauty of quantum mechanics. At the very fundamental level, the reader has to realize quantum mechanics as a model of reality.

The French philosopher - mathematician Henri Poincare said, "it is hardly necessary to point out how much quantum theory deviates from everything that one has imagined until now; it is, without doubt, the greatest and the deepest revolution to which natural philosophy has been subjected to since Newton". Truly, it is very difficult to accept such a revolution, which is quite contrary to everyday experience. The main trouble in learning quantum mechanics is that the mind will not be ready to accept the facts connected with situations unfamiliar to us. This can be achieved by gaining "QUANTUM INSIGHT" into the nature of reality and such an insight will allow us to think about the universe in a different perspective.

Unfortunately, most of the students and even the teachers are reluctant to look into the text books of quantum mechanics, simply because of the discomfort resulting out of the diffidence developed over the years. In view of this fact, an attempt made to bring out a book on Quantum Chemistry, to provide the reader with necessary background to venture into the realms of higher quantum mechanics. To make the book more appealing and interesting little chunks of history, philosophy and biographies are included at appropriate places in the text. I consider my effort be rewarded if it can make the average student realize and extract the beauty of quantum mechanics from its abstract laws and equations, applicable not just to the atomic domain, but to the real world as well.

- Author

Contents

Preface .... (vii)

CHAPTER 1
Historical Background

1.1 Newtonian Mechanics 1
1.2 Black Body Radiation 3
        Physical Basis for the Success 6
1.3 The Photoelectric Effect 7
1.4 The Compton Effect 9
1.5 Atomic Spectra 10
1.6 Atomic Models 11
1.7 The Bohr Atom 13
        Assumptions 13
        Energy of the Electron in the Atom 14
        Extensions of the Bohr's Theory 15
        Zeeman Effect 16
        Spin 17
1.8 Failure of the Old Quantum Theory 17

CHAPTER 2
The Wave Equation

2.1 de Broglie's Concept of Matter Waves 19
        Wave length and momentum of a particle 20
2.2 Heisenberg's Uncertainty Principle 22
2.3 Wave Equation 26
2.4 Interpretation of Wave Function 29
2.5 Normalized and Orthogonal Wave Functions 30
2.6 Exercises 31

CHAPTER 3
The Postulates

- 3.1 The Formulation of Quantum Mechanics 33
  - Schrödinger Wave Equation 33
- 3.2 The Postulates of Quantum Mechanics 34
  - 3.2.1 Postulate I 34
  - 3.2.2 Well Behaved Wave Function 35
    - The Fitness of the Wave Function 36
  - 3.2.3 Postulate II 37
    - Hermitian Operator 38
  - 3.2.4 Postulate III 41
  - 3.2.5 Postulate IV 42
  - 3.2.6 Postulate V 43
- 3.3 Exercises 43

CHAPTER 4
Applications of Schrödinger Equation-1
(Simple systems with constant potential energy)

- 4.1 Particle in a One-dimensional Box 48
  - 4.1.1 Salient Instructive Features of the Problem 51
  - 4.1.2 Zero Point Energy 55
  - 4.1.3 Free Particle 56
- 4.2 The Particle in a Three Dimensional Box 56
  - Degeneracy 59
- 4.3 The Structure of Matter 62
- 4.4 Factors Influencing Color 64
- 4.5 Tunneling in Quantum Mechanics 67
  - Systems with Discontinuity in the Potential Field 69
  - Hydrogen Transfer Reaction 71
- 4.6 The Rigid Rotor 72

CHAPTER 5

Applications of Schrödinger Equation-2
(Simple Systems with Variable Potential Energy)

- 5.1 One-dimensional Harmonic Oscillator 77
  - Wave functions of the harmonic oscillator 80
- 5.2 The Hydrogen Atom 82
  - Polar coordinates 83
  - Separation of variables 85
  - The ϕ equation 88
  - The θ equation 89
  - Spherical Harmonics 90
  - The Radial Equation 90
  - Quantum States 91
  - Wave Functions of the Hydrogen Atom 92
  - Hydrogen like Wave Functions 94
  - The Radial Function 94
  - The Radial Distribution Functions 95
  - Show that $r = a_0$ for the 1s orbital 97
  - The Angular Function $Y_{(\theta,\phi)}$ 98
  - Nomenclature of p Orbitals 99
  - Nomenclature of d Orbitals 100

CHAPTER 6

Approximation Methods

- 6.1 Perturbation Theory 101
  - Perturbation theory consists of a set of successive corrections to an unperturbed problem 103
- 6.2 The Variational Method 105
  - Proof 107
- 6.3 The Hartree Theory 108
  - Surface of the Atom 108
- 6.4 Exercises 112

CHAPTER 7
Bonding in Molecules

- 7.1 Molecular Orbital Theory 115
  - Hamiltonian operator for H_2^+ and H_2 123
  - The Stability of Hydrogen Molecule Ion 124
- 7.2 Valence Bond Theory 127
- 7.3 Hybridization 131
  - Linear Structure – $BeCl_2$ 132
  - Trigonal Planar Structure 134
  - Tetrahedral Structure 135
  - Octahedral Complexes 137

CHAPTER 8
Appendix

- 8.1 SI Units (Système International d'unités) 139
- 8.2 Derived Units 139
- 8.3 Supplementary Units 141
- 8.4 CGS Units 141
- 8.5 Prefix Dictionary 141
- 8.6 Experimental Foundation 142
- 8.7 Calculation of Effective Nuclear Charge 143
- 8.8 Approximate Orbitals 146
  - Slater orbitals 146
- 8.9 Angular Momentum 147
- 8.10 Laplacian Operator 150
  - (Conversion from Cartesian to Polar coordinates) 150
- 8.11 Supplement to Rigid Rotor 151
  - Associated Legendre function 151
  - Associated Legendre polynomial 151
- 8.12 Supplement to One-dimensional Harmonic Oscillator . 151

CHAPTER 1

Historical Background

1.1 Newtonian Mechanics

Newtonian mechanics or classical mechanics in its simplest form, known as the laws of mechanics is written in terms of particle trajectories. In fact, the trajectory underlies the structure of classical physics and the particle underlies the model of physical reality. The underlying assumptions and philosophical implications of classical physics are so familiar that we have never given them a second thought. Classical physics ascribes to the universe an Objective Reality, an existence external to and independent of human observers.

Our central assumption about the nature of classical universe is that, it is predictable. Knowing the initial conditions of a system, however complicated it might be, we can use Newton's laws to predict its future. This notion is the essence of determinism that supported Newtonian mechanics for more than three centuries.

Newtonian mechanics has taken such strong roots and everybody believed that everything in this universe can be explained on the basis of these laws. Many scientists have predicted the end of science as they thought that there is nothing new to know and nothing more to investigate. In fact, Prof. John Trowbridge at Harvard University, the then Head of the Department, felt compelled to warn bright students away from physics. He told them that the essential business of Science is over. All that remains is to dot a few 'i's and cross a few 't's, a task best left to second rate.

In 1994, Albert Michelson, the future recipient of the Noble Prize told the audience in one of the conferences that "it seems probable that most of the underlying principles have been firmly established and that further advances are to be sought chiefly in the rigorous application of these principles to all phenomena which come under our notice. The future truths of physics are only to be looked for, in the sixth place of decimals".

However, these ideas did not long lost and with the discovery of x-rays, radioactivity and electron in the last decade of the 19th century, the scientists had to think afresh about the universe.

Röentgen discovered x-rays in his laboratory at Wurtzburg in 1895. For this discovery, he received Rumford medal of the Royal Society in 1896 and the first Nobel Prize in physics in 1901. Henry Bacquerel in 1896 trying to reproduce Röentgen's x-rays, accidentally discovered radioactivity in potassium uranyl sulphate, a phosphorescent rock available in his laboratory. For this discovery, he shared the 1903 Noble prize in physics with the Curies.

In 1897, the British Physicist J. J. Thompson demonstrated that the beam that leaves the cathode, the so-called cathode rays, consists of a beam of negatively charged discrete particles. By balancing this beam between an electric and magnetic field, Thompson was able to measure the charge to mass ratio of these particles, the currently accepted value being 1.7588×10^{11} Coulombs/Kilogram (C.Kg^{-1}). Thompson also estimated the charge on the electron by utilizing the observation by C. T. R. Wilson (of the Wilson cloud chamber) that a charged particle acts as nucleus around which water vapour condenses. Thus by performing an early version of the famous oil drop experiment of Millikan, Thompson calculated the charge on the electron to be about 1×10^{-19} C and its mass to be about 6×10^{-31} Kg. Although Thompson's charge to mass measurement was quite accurate, his determination of charge itself was in error by 50%. Consequently, his calculation of the electronic mass was in error by 50%. Nevertheless, he did show that an electron was much lighter than the lightest atom and so it should be a subatomic particle. A little over 10 years later, Millikan refined the electron charge as 1.60×10^{-19} C almost getting the modern value of 1.6022×10^{-19} C.

Although these experiments did not lead immediately to the realization of the inadequacy of the classical physics, they showed that the atom was far more complex than had previously been thought. It was a major challenge to classical physics to provide a structure for the atom, but this was a challenge to which classical physics never rose.

J. J. Thompson
(1856-1940)

Thompson Studied Engineering at Owens College where he developed interest in Science. In 1876, he went to Cambridge University on a scholarship and remained there for the rest of his life. In 1884, he succeeded Lord Rayleigh as the Cavendish Professor of Physics and Director of the Cavendish laboratory. Thompson was an excellent teacher and administrator. Seven Nobel Prize Winners were trained under Thompson at the Cavendish. In 1919, he resigned his Directorship in favour of Ernest Rutherford, in part because of his lack of sympathy for the new Physics of Niels Bohr. Thompson was awarded the Nobel Prize in Physics in 1906 and was knighted in 1908.

1.2 Black Body Radiation

When a body is heated it emits thermal radiation, and the nature of this radiation depends on the temperature of the emitting body. When the heating element of an electric stove is turned on, it emits radiation. This radiation can be detected by placing one's hand at some distance above the heating element. If the stove is on low heat, the radiation can be detected by feeling only and not by sight. If the heat is turned up, the stove element will begin to glow first red, then white and if the temperature could be raised high enough, even blue. This change in colour is evidence that the frequency distribution of the radiation emitted by the hot body is changing with temperature.

In order to study such radiation, it was found that a particularly desirable system was one known as a "black body". When radiation falls on a surface, some of the radiation is reflected and some is absorbed. The absorptivity of a surface is defined as the fraction of the light incident on the surface that is absorbed, and a black body is defined as one that has an absorptivity of unity. That is, it absorbs all the radiation that is incident upon it. In addition, it has been shown (Kirchhoff's law) that the ratio of emissive power, 'E', to the absorptivity, 'A' i.e.

$$\frac{E_0}{A_0} = \frac{E}{A}$$

is a constant for a given temperature.

Now, since the absorptivity of a black body has been defined as unity ($A_0 = 1$), we see that the total emissive power of any surface must be given by, $E = AE_0$ where E_0 = total emissive power of a black body. Since 'A' is necessarily less than unity for any surface other than a black body, it is obvious that no surface can emit more strongly than a black body. *Therefore, it is seen that a black body is both the most efficient absorber and also the most efficient emitter of radiant energy.*

Many experiments were carried out on the black body radiation. The apparatus used for the study of black body radiation consists of a well insulated cavity with a small opening in one of the walls, and this type of furnace is kept at constant temperature. This furnace is called an isothermal enclosure and the radiation is observed as it passes through the small hole or opening. In 1858, Kirchhoff was able to show that if the walls and contents of the cavity are kept at a constant temperature at equilibrium, the stream of radiation in one direction must be the same as that in any other direction. It must be the same at any point in the enclosure and makes no difference of what material the walls are composed.

In 1879, Stefan had given an empirical relation for the rate of emission of radiant energy per unit area of a surface. *(The law was experimentally discovered by Stefan in 1879 and derived by Boltzman in 1884 based on the principles of thermodynamics)*

$$E = e\sigma T^4$$

where E = Rate of emission of radiant energy per unit area, (or the total emissive power), T = Absolute temperature, e = emissivity of the surface, σ = Stefan-Boltzman constant. (Emissivity is defined as E/E_0 and for a black body emissivity, $E_0 = 1$).

A problem that was of considerable interest at that time was the distribution of energy in the spectrum as a function of wavelength and temperature. In 1894, Willy Wien has provided another useful piece of information in the form of displacement law. It says that the wavelength that corresponded to the maximum of the energy distribution of the black-body radiation, obeys the relation,

$$\lambda_{max} T = \text{Constant}.$$

This is a consequence of the theoretical attempts made to calculate the shapes of the energy spectra as a function of wavelength.

In an attempt to find an expression for the monochromatic emissive power, Wien utilised the classical methods of thermodynamics to obtain the equation

$$E_\lambda = \frac{a}{\lambda^5} f(\lambda T)$$

where 'a' is constant and $f(\lambda T)$ is a function of λ & T.

In order to determine the function $f(\lambda T)$, it was necessary to consider the mechanism by which the radiation is emitted. Since, Kirchhoff had shown that the nature of the walls, and therefore the nature of the radiator, is not important in an isothermal enclosure, any reasonable model can be chosen. Wien chose oscillators of molecular size and applied the laws of classical electromagnetic theory. He obtained the equation,

$$E_\lambda = \frac{a}{\lambda^5} e^{-b/\lambda T}$$

where 'a' and 'b' are constants.

Another theoretical attempt to determine a distribution law was made in 1900 by Rayleigh, by applying the equipartition principle to electromagnetic field. This calculation consists of two parts. In the first, one calculates the number of oscillators in an enclosure that correspond to a wavelength, λ. The second part, in accord with the classical equipartition principle, involves associating an energy, KT with each oscillator. Jean commented on some of the mistakes in the calculation and their combined effort, resulted in the form of a modified equation, known as Rayleigh-Jeans equation,

$$E_\lambda = \frac{2\pi kT}{c\lambda^4}$$

Almost simultaneously in 1899, Lummer and Pringsheim made the experimental determination of the energy distribution from a black body at various values of the temperature. The results are shown in Fig. 1.2.1.

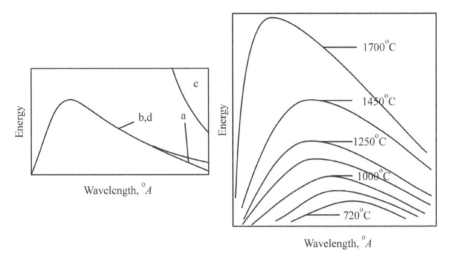

Fig. 1.2.1 Comparison of the three radiation laws with the experimental data.
(a) Wien (b) Planck and (c) Rayleigh-Jeans with (d) the dotted experimental curve.

Wien equation gives excellent agreement with experiment in the region of short wavelengths and the Rayleigh-Jeans equation appears to be asymptotically correct at long wavelengths. This equation is clearly not correct, since, it predicts an impossible situation, namely, at shorter and shorter wavelengths the radiation intensity should increase without bound. This paradox known as the *"ultraviolet catastrophe"*, dealt a terrible blow to the 19th century classical physics. Hence, neither of the equations is consistent with the experimental curves over the complete spectral range.

Many attempts were made to propose equations to fit into the total experimental spectrum. Such an attempt by Max Planck has brought out the most revolutionary hypothesis of the era.

For the same reason, Wien was able to choose any type of energy radiator that he wished, Planck too made such a choice. It had to be a system capable of emitting and absorbing radiation, and among those the simplest type for the purpose of calculation is a set of simple harmonic oscillators. Now, according to classical ideas, an oscillator must take up energy continuously and emit energy continuously. However, in order to find a formula that would fit the experimentally determined spectrum of a black body radiator, Planck found it necessary to postulate that such an oscillator cannot take up energy continuously as demanded by classical theory, but rather it must take energy in discrete amounts.

These amounts are integral multiples of a fundamental energy unit ε_o that is, 0, ε_o, $2\varepsilon_o$, ... $n\varepsilon_o$.

Using this idea, Planck was able to derive the equation,

$$E_\lambda = \frac{2\pi c}{\lambda^4} \frac{\varepsilon_o}{e^{\varepsilon_o/kT} - 1}$$

for the monochromatic emissive power of a black body. Here, 'c' is the velocity of light and 'k' is the Boltzman constant. Since the Wien equation is of thermodynamic origin, and therefore correct, it is necessary for the distribution law of Planck to contain the temperature in the combination, T or (T/v) or (v/T).

Consequently, it can be seen that the quantum of energy, ε_o must be proportional to $1/\lambda$ or v. We find that $\varepsilon_o = h\nu$, where h = Planck's constant. By making the substitution for ε_o.

$$E_\lambda = \frac{2\pi c^2}{\lambda^5} \frac{1}{e^{ch/\lambda kT} - 1}$$

Whereas the energy distribution laws for black body radiation deduced from classical concepts had consistently failed to explain the experimental data, the quantum hypothesis of Planck succeeded. The hypothesis involves no extension of classical ideas, but it is a radical change from the prevalent line of thought of that time. Quite in contrast to the classical idea that an oscillator can absorb and emit energy continuously from wavelengths of zero to infinity, Planck proposed that the energy must be emitted or absorbed, only in discrete amounts. This implies that any system capable of emitting radiation must have a set of energy states, and emission can take place only when the system changes from one of these energy states to another. Intermediate energy states do not occur. Thus, we may find an oscillator emitting an energy of 2hv, but not 0.5 hv.

Physical Basis for the Success

The physical basis for the success of the quantum hypothesis may be, due to the fact that, at a particular temperature there may not be sufficient energy available to excite the higher frequency oscillators. It is because, based on the quantum hypothesis, they can be excited only by absorbing not less than one quantum of energy, $h\nu$. On the classical theory, the oscillators could be excited in a continuous manner. Therefore, at the temperature T, when the mean thermal energy available is kT, even the highest frequency oscillators could be excited with a frequency 'ν' (and by equipartition, an energy kT) and so contribute to the radiation from the emitter. Planck's quantum hypothesis therefore has the effect of damping out the high-frequency oscillators, just as we realised was necessary.

Black body radiation is a fundamental problem, and we have arrived at a solution by making a radical alteration to classical theory. Therefore, we should expect to discover ramifications of hypothesis in other parts of physics and chemistry.

It was not long before Planck's hypothesis had another application. In 1905, in order to explain photoelectric effect, Albert Einstein postulated that light energy had to be quantized.

1.3 The Photoelectric Effect

In 1887, Hertz observed photoelectric effect and the first outstanding application of quantum theory was in its explanation by Albert Einstein in 1905. It should be noted that though Planck introduced the idea that radiation must be emitted in quanta or bundles of energy, he however believed that, after being so emitted, the radiation spread in waves. Einstein extended Planck's idea further and introduced the important concept that the radiation energy is not only emitted in quanta but the quanta also preserved their identity until they were finally absorbed.

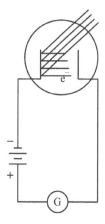

Fig. 1.3.1 Schematic diagram of apparatus for investigating the photoelectric effect.

Photoelectric effect is the ejection of electrons from various materials when irradiated by visible or ultraviolet light. This effect is the basis of photoelectric cell, an extremely sensitive instrument used for detection and measurement of radiation. An arrangement that can be used for this study is shown in Fig.1.3.1.

Laws of photoelectricity, established from experimental facts are as follows:
(a) The total photoelectric current is proportional to the intensity of the light striking the surface.
(b) For each particular metal used to form the surface, there exists a threshold frequency or (wavelength) such that, at frequencies below the threshold, no electrons are emitted, no matter how great the intensity might be.

(c) The maximum energy of the emitted electrons is independent of the light intensity.

(d) The maximum energy of the emitted electrons is linearly dependent on the frequency of the incident light. Fig.1.3.2.

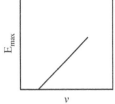

Item (a) is quite expected, item (b) involving discontinuity is surprising, item (c) is totally unexpected and item (d) is an unexplained phenomenon.

Fig. 1.3.2 Variation of the maximum energy of the photoelectrons with frequency of the incident radiation.

Clearly, the photoelectric effect must require an explanation radically different from classical electromagnetic theory. Einstein's celebrated note of 1905 provided the correct explanation. Going even further than Planck, who limited himself to the introduction of discontinuity in the mechanism of absorption and emission, Einstein postulated that light radiation itself was discontinuous, consisting of beam of corpuscles, named as photons. A photon is thus a single quantum of electromagnetic radiation and has the energy, $h\nu$.

According to the Einstein's explanation, when a photon strikes a metal surface, a given electron on the surface would receive either all of its energy '$h\nu$' or no energy at all. Again as the electron escapes from the metal, it uses up certain energy, W, in overcoming the surface forces, called the work function. Moreover, if the electron originates below the surface, additional amount of energy may be used up in reaching the surface. So, for an electron originating at the surface or one which loses no energy in reaching the surface, the Kinetic Energy (K.E.) after leaving the surface will be the maximum.

Obviously, this K.E. is the difference between $h\nu$ and W.

So $(½) mv^2 = h\nu - w$

$= h\nu - h\nu_0$

So $(½) mv^2_{max}$ = stopping potential = $h\nu - w$ or $h\nu - h\nu_0$.

Thus, we can see that if the energy of the incident photon is less than the energy needed by the electron to escape from the surface, no emission can take place, regardless of the intensity of the incident light, i.e., the number of photons, which strike the surface per second.

According to classical physics, electromagnetic radiation is an electric field oscillating perpendicular to its direction of propagation, and the intensity of the radiation is proportional to the square of the amplitude of the electric field. As the intensity increases, so does the amplitude of the oscillating electric field.

The electrons at the surface of the metal oscillate along with the field. As the intensity (amplitude) of the field increases, the electrons oscillate more violently and eventually break away from the surface with a kinetic energy that depends on the amplitude (intensity) of the field. This nice classical picture is at complete variance with the experimental observations. Further, this classical picture predicts that the photoelectric effect should occur for any frequency of the light as long as the intensity is sufficiently intense.

1.4 The Compton Effect

The Compton effect provided further evidence for the quantum nature of radiation. If photons are really particles, they should possess a momentum, 'p', equivalent to hv/c. This momentum should be observable by allowing a beam of light to fall on a beam of electrons, when a transfer of momentum should be observed as a scattering of the electrons by the light beam, or as a scattering of the photons by the electrons. Compton performed this experiment using x-rays as the light beam in 1922, and the results he obtained were in complete agreement with the predictions.

It has been observed that when monochromatic X-rays impinge on elements of low atomic weight, the scattered X-rays were found to be of longer wavelength than those of the impinging beam. This phenomenon cannot be explained based on the classical theory, because as per this theory monochromatic light falling upon matter should be scattered without change in frequency.

However, the effect could be satisfactorily explained as resulting from an impact between the X-ray photon and the electron. Because of this collision, the electron recoils and the photon is scattered. In the process the electron gains momentum and the photon loses momentum. The decrease in momentum of the photon is manifested in the form of lowering of its frequency or increase in its wavelength.

It has also been shown that only one value of the wavelength shift is observed at a given scattering angle: this implies that the momentum transfer takes place only in a discrete manner and not continuously.

The photoelectric effect is stronger than Compton effect when X-rays of energy less than 0.1 MeV are used. In the process of the photoelectric effect, the energy of an X-ray photon is completely given up to an electron of the atomic system. Since it is impossible for a photon to give up all its energy to a free electron, the photoelectric effect can take place only when photons strike bound electrons. At higher X-ray energies (about 0.1 MeV) the Compton Scattering becomes more important. In this case, an X-ray photon is scattered and not really absorbed, since it does not lose a very large fraction of its energy. At still higher energies, above 1 MeV (wavelengths less than

0.0120 Å), the process of pair formation plays a part in the absorption of X-rays. As the photon energy increases this process becomes more important than either photoelectric absorption or Compton Scattering.

Gamma rays are very short electromagnetic waves whose energy range overlaps that of X-rays and extends to several MeV. Under suitable conditions, a gamma-ray-photon converts itself into a pair of material particles, a positive electron and a negative electron. The former is called the positron, while the word electron is used only for negative electrons. In this process of pair formation discovered by Anderson in 1932, two particles each of mass m_0 are created out of the energy of the gamma-ray, if the initial energy is at least equal to $2m_0c^2$. According to Einstein's mass-energy relation $E = mc^2$, the energy required to create an electron is 0.511 MeV. Thus, pair formation cannot take place until the energy of the photon is at least 1.022 MeV the threshold energy for pair formation. Experimentally also this is found to be true. If the initial photon energy is greater than this threshold value, the excess appears as kinetic energy shared equally by the positron and the electron.

1.5 Atomic Spectra

At a time when people were engaged with the problems of black body radiation, a similar development was taking place in the field of atomic spectra.
1. *It was observed that when an electric discharge is passed through an element in the gaseous state, light will be emitted*
2. *Analysis of this light by a prism or grating spectrometer gives a series of sharp lines of a definite wavelength, which prove to be characteristic of the particular element.*

In the case of light element such as hydrogen, this line spectrum turns out to be simple, but for heavier elements, it is more likely to be extremely complex. As the experimental data accumulated, people observed some sort of orderliness, and so, tried to obtain empirical relations to predict the sequence of lines.
1. In 1883, Liveing and Dewar realised that several possible series exist in the spectra of alkali and alkaline earth metals but could not discover an empirical relation to present the order.
2. In 1885, Balmer discovered the equation.

$$\lambda = \frac{bn^2}{n^2 - 4}$$

where 'b' is a numerical constant and 'n' is an integer, e.g. 3, 4, 5......etc. The agreement between the observed values, of the lines in the hydrogen spectrum and their values calculated by the Balmer formula, turns out to be extremely good.

$$v = \frac{c(n^2-4)}{bn^2} = Rc\left(\frac{1}{2^2} - \frac{1}{n^2}\right)$$

The above equation can be expressed as

$$\bar{v} = R\left(\frac{1}{2^2} - \frac{1}{n^2}\right)$$

where R = Rydberg constant.

The Rydberg constant has been found to be specific for a given element and very nearly constant for all elements. The difference in its value is due to the atomic weight of the element, and it has been found to have a value of 109,677.58 cm^{-1} for hydrogen.

At the time when the Balmer series was discovered, the known portion of the electromagnetic spectrum was the visible region (4000 to 8000Å) alone. After this discovery, the same general type of other series were discovered. The Lymann series was found in the ultraviolet region and the Paschen, Brackett and Pfund series were found in the infrared.

The general equation can be written as:

$$\bar{v} = R\left(\frac{1}{n_1^2} - \frac{1}{n_2^2}\right)$$

where
- $n_1 = 1$ Lyman series   u.v.
- $n_1 = 2$ Balmer series   Visible
- $n_1 = 3$ Paschen series   Near IR
- $n_1 = 4$ Brackett series Far IR
- $n_1 = 5$ Pfund series   Far IR

and $n_2 > n_1$

Although the early developments in atomic spectra were significant, they were nevertheless empirical. For the most part, they were restricted to classifying and correlating the observed data by means of empirical relations, and there was no clue how these spectral lines arouse.

1.6 Atomic Models

The origin for the spectral lines could be the atoms, is a reasonable assumption. But, how the atoms are able to emit such characteristic lines has remained a matter of speculation because of the absence of any satisfactory concept of the structure of the atom. Subsequently things became clear.

12   **Quantum Chemistry**

1. With the discovery of radioactivity and the emission of positive, negative and a number of combinations of these particles, it became clear that atom is composed of these newly found particles. So immediately the next question will be, how many of each category are there and how they are arranged in an atom.

2. Basing on the available data at that time, J. J. Thompson proposed a model of the atom with the positive charge distributed uniformly throughout a sphere of diameter 10^{-8} cm. The electrons are embedded in the sphere in equilibrium positions and when disturbed, they oscillate about these equilibrium positions.

   Though it is a crude model, it could account for the occurrence of spectral lines, but it could not explain the scattering of 'α' particles ($_2^4He^{2+}$).

   One of the ways by means of which these 'α' particles can be observed is by the scintillations they cause on a fluorescent screen coated with zinc sulphide. When a thin gold foil is placed in the path of the 'α' particles, naturally a change in pattern on the screen is expected, compared to the one obtained without the gold foil in the path. However, the immediate question will be "How it will change?" Therefore, Thompson calculated theoretically and concluded that the average deflection of the 'α' particles should be small and the probability of the large scale scattering is essentially zero. But, Geiger and Marsden noted experimentally that about 1 in 8000 'α' particles, incident on a gold foil, is deflected through an angle greater than 900. This is in complete disagreement with Thompson's model and his predictions.

3. To resolve this, Rutherford proposed a new model of an atom in which the positive charge is concentrated in a small volume at the centre of the atom. The electrons are then assumed to move around this centre of positive charge in various orbits, as the planets in the solar system. This is an improvement over the Thompson's model as it accounts for the wide angle scattering of the 'α' particles in the gold leaf experiment. However, it also met with some difficulties.

   (a) The electrons could not be considered to be stationary because the unlike charges of the electron and the nucleus cause them to come together.

   (b) If the electrons are assumed to be moving around the nucleus, another problem arises. When an electric charge is accelerated, it emits or absorbs radiation. If the electrons are pictured as moving around the nucleus, they are subject to centripetal acceleration. According to the principles of electromagnetic theory, the electrons therefore must radiate energy. The only place for this continuous supply of energy is the atom itself, and eventually the electron should spiral into the nucleus and in essence run down. Hence, Rutherford's model is not the final answer.

1.7 The Bohr Atom

Through many models were proposed, the model proposed by Niels Bohr (1913) for the hydrogen like atom, is unique in gaining universal recognition. Using the structural ideas of the Rutherford atom, Bohr was successful in quantitatively applying the concepts of quantum theory to explain the origin of line spectra as well as the stability of the atom.

Bohr was able to overcome the difficulties encountered in the earlier model, by applying the quantum concept of discrete energy states.

Assumptions

1. The electron in an atom is restricted to move in a particular stable orbit, and as long as it remains in this orbit, it will not radiate energy.
2. Using the quantum principle, that an oscillator will not emit energy except when it jumps from one energy state to another, Bohr postulated that when the electron jumps from a stable energy state of energy E_1 to another state of lower energy E_2, a quantum of radiation is emitted, with an energy equal to the difference between the two states.

$$h\nu = E_1 - E_2$$

3. In the final form of the theory, Bohr assumed the orbits to be circular with a size such as to satisfy the quantum condition that the angular momentum, p, of the electron is an integral multiple of the quantity $h/2\pi$. Thus

$$p = \frac{nh}{2\pi} = mvr.$$

$$\therefore \quad v = \frac{nh}{2\pi mr}$$

where m and v are the mass and velocity of the electron. 'h' is the Planck's constant and 'n' is a positive integer known as a quantum number.

For a quantitative treatment of a one electron system, the force of attraction between the electron and the nucleus is considered to arise from the electrostatic attraction between the positive charge of the nucleus and the negative charge of the electron, thus $F = Ze^2/r^2$, where Z is the atomic number of the element and 'r' is the distance between the nucleus and the electron.

This electrostatic attraction should be equal to the centripetal force resulting from the motion of the electron about the nucleus

14 Quantum Chemistry

$$\therefore \quad \frac{Ze^2}{r^2} = \frac{mv^2}{r}$$

$$\therefore \quad r = \frac{Ze^2}{mv^2}$$

But according to the quantum condition

$$p = n \cdot \frac{h}{2\pi} = mvr$$

Hence, substituting the value of 'v' in the above equation

$$r = \frac{n^2 h^2}{4\pi^2 \, mze^2}$$

for the hydrogen atom, Z = 1 and if the electron is considered to be in the ground state (n = 1) the radius of the atom can readily be calculated to be r = 0.529 A⁰.

Energy of the Electron in the Atom

The total energy of the electron of the atom is made up of its kinetic and potential energies. If zero of potential energy is defined as the energy of the electron when it is at rest at an infinite distance from the nucleus, its potential energy with respect to the nucleus at any distance 'r' is found to be

$$v = \int_\alpha^r F \cdot dr = \int_\alpha^r \frac{Ze^2}{r^2} dr = \frac{-Ze^2}{r}$$

The kinetic energy, $T = \frac{1}{2} mu^2 = \frac{Ze^2}{2r}$

$\therefore$ Total energy of the electron = T + V

$$= \frac{-Ze^2}{r} + \frac{Ze^2}{2r} = \frac{-Ze^2}{2r}$$

$\therefore$ The energy of the electron in the nth quantum state

$$E_n = \frac{-ze^2}{2r_n} = \frac{-ze^2 \times 4\pi^2 mze^2}{2n^2 h^2}$$

$$= \frac{-2\pi^2 me^4 z^2}{n^2 h^2}$$

Thus the transition between two energy states of energy E_1 and E_2 an be written as

$$\nu = \frac{E_2 - E_1}{h} \quad \text{or} \quad \bar{\nu} = \frac{2\pi^2 me^4 z^2}{ch^3}\left(\frac{1}{n_1^2} - \frac{1}{n_2^2}\right)$$

If $n_2 = 2$, it is exactly same as the Balmer equation. The constant term in the above equation has given a reasonable agreement with the Rydberg constant. This gave overwhelming support to Bohr's theory.

Now that it has been shown that the equation for the wave number developed by Bohr is same as that found by Balmer, it is now possible to explain the origin of spectral series.

In the equation

$$\bar{\nu} = R\left(\frac{1}{n_1^2} - \frac{1}{n_2^2}\right)$$

$n_1 = 2$ arises from the fact that the electron transitions are to the second shell. In a similar manner, an analogous relation exists between $n_1 = 1$ and the Lymann series and $n_1 = 3$, 4 and 5 for the Paschen, Brackett and Pfund series, respectively.

Extensions of the Bohr's Theory

Even though Bohr's theory could predict the energies of the spectral lines of the hydrogen like atoms, it met with some difficulties also.

1. It could not explain the fine structure in the line spectrum of the hydrogen-like atom. When Bohr proposed the theory, only single lines were observed and the theory successfully predicted them. But as better instruments and techniques are developed, it was realised that what were thought to be single lines, were actually a collection of several lines close together. This implies that there are several energy levels close together rather than a single level for each quantum number 'n'. This would require new quantum numbers and there is no way to obtain them directly from Bohr model.

   This problem was solved by Sommerfeld when he considered in detail the effect of elliptical orbits for the electron. For an elliptical orbit, both the angle 'ϕ' and the radius vector 'r' can vary.

   Summerfeld found that the degeneracy in this atomic model can be removed by considering the relativistic change in the mass of the electron during its motion around the nucleus. As the electron revolves around the nucleus, its velocity changes continuously, depending on its proximity to the nucleus. From the special theory of relativity it is known that the mass of a particle increases as its velocity increases. If this effect is taken into consideration, a small difference in energy is found to exist between a circular orbit and an elliptical orbit. This difference is a function of '$n\phi$', and can be related to the physical picture of energy level in the Bohr atom by considering each major energy level to be composed of several sub-levels lying very close together.

The change in mass of the electron produces slight changes in the effective columbic forces operating between the electron and the nucleus. If this effect is taken into consideration, a small difference in energy is found to exist between orbits of different eccentricities, which will be reflected as fine structure in the spectrum.

The explanation of the fine structure of the spectrum of hydrogen is a notable achievement of the Sommerfeld's modification. But, the greatest single contribution of the Sommerfeld's concept, however, lies in the subdivision of the original Bohr stationary states into several sub-states of slightly differing energies as characterized by orbits of different eccentricities. Inherent also in the concept of elliptical orbits is the concept of penetrating orbits. We shall see later that these features form the basis to the modern concepts of electronic configuration.

Zeeman Effect

When the source emitting the spectral lines was placed in a strong magnetic field, a further splitting of lines was noticed. In order to account for this phenomenon called the Zeeman effect, a third quantum number known as the magnetic quantum number was postulated.

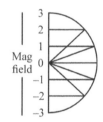

Fig. 1.7.1 Space quantization in a magnetic field.

An electron in space requires three coordinates to describe its position. This has three degrees of freedom and should require three quantum numbers to describe its energy. Without a spatial reference, the arrangement of the orbital plane of the electron is completely arbitrary, and this third degree of freedom is degenerate.

However, in the presence of an external field, the orbital plane of the electron will precess about the direction of the field, and thereby remove the degeneracy. The possible positions the orbital plane (Vector representing the orbital angular momentum) can assume in space are limited (Space quantisation) and the magnitude of its component in the direction of the magnetic field is given in terms of the magnetic quantum number 'm'.

The third quantum condition, similar to that of the angular momentum is

$$P_z = m \frac{h}{2\pi}$$

The magnetic quantum number may have any integral value including zero from $-l$ to $+l$ giving a total of $2l + 1$ values.

The possible values of m when $l = 3$ are shown in Fig.1.7.1. Positive values of 'm' describe the components of orbital angular momentum oriented in the direction of the applied magnetic field, and the negative values represent those oriented in the opposite direction.

Spin

The presence of 'double lines' in the spectra of alkali metals was attributed to the axial spin of the electron by Goudsmit and Uhlenbeck (1925).

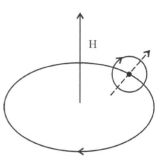

A simplified account of the way in which this property leads to new energy levels can be understood, if we remember that a spinning electron behaves as a small magnet. Now, the electron moving around in its orbit produces a magnetic field just as an electric current in a coil of wire produces a field. The arrow marked 'H' represents this field. Since the electron, because of the axial spin behaves as a small magnet, there will be an interaction between the two magnetic fields. The field produced by the axial spin either reinforces or opposes the field 'H' depending on the direction of the spin whether it is clockwise or anticlockwise.

This interaction will produce energy changes, wherein, a single energy level representing a non-spinning electron moving in an orbit, becomes two energy levels close together. Additional electron transitions are therefore possible, and new lines appear in the spectrum. Goudsmit and Uhlenbeck showed that the spectroscopic observations required that the angular momentum associated with the spin of the electron is given by $m_s.h/2\pi$, where m_s is called the 'Spin quantum number' and can have the value $+1/2$ or $-1/2$.

1.8 Failure of the Old Quantum Theory

The success achieved by the Bohr-Sommerfeld theory in explaining the atomic spectrum of hydrogen prompted its extension to other systems. Although it achieved some success in accounting for the spectra of such hydrogen like species as singly ionized helium (He^+), doubly ionized lithium (Li^{2+}) and triply ionized beryllium (Be^{3+}), it failed to predict the spectral lines and spectral intensities in the case of many electron atoms. Apart from this, there are certain other unsatisfactory features in the theory. For instance, there is no justification for the assumption that an electron can move in only those orbits wherein the angular momentum of the electron is an integral multiple of h/2π. Further whenever it has become necessary to explain an experimental observation a new quantum number has been introduced; thus, the introduction of the various quantum numbers is arbitrary.

Finally, the theory contributed virtually nothing to an understanding of the geometry of the molecules.

The unsatisfactory features in the old quantum theory led scientists to search for new mechanics for the treatment of atomic systems that would relieve the wave particle conflict and introduce quantized energy. As a consequence of some more efforts in this direction culminated in the formulation of 'matrix mechanics' by Heisenberg and 'Wave mechanics' by Schrödinger. Afterwards, Schrödinger and Eckart have shown that both matrix mechanics and wave mechanics lead to the same conclusions. Now these two forms of mechanics are covered by the term 'quantum mechanics': that means matrix mechanics and wave mechanics are merely two different mathematical treatments of quantum mechanics. Of these two forms, since wave mechanics are easy to understand and can explain all the phenomena in chemistry, this method is widely used in quantum chemistry.

Quantum mechanics, not only could explain the phenomena associated with chemistry, it helped to amalgamate physics, chemistry and material science. Earlier also the attempts of Newton were successful in applying common laws to the celestial and terrestrial objects alike. Similarly, Mayer and others have unified the laws of heat and mechanics, while Faraday and Maxwell have shown that electricity, magnetism and optics are closely related. Einstein was responsible for bringing together space, time, matter and gravity. The scientific community is eagerly awaiting for a theory which can explain everything in the atomic, nuclear and sub nuclear levels and beyond that includes the bigger than the biggest and the smaller than the smallest, known to us at this point of time.

CHAPTER 2

The Wave Equation

2.1 de Broglie's Concept of Matter Waves

It has already been pointed out that electromagnetic radiation exhibits a dualistic character. Under certain experimental conditions, it is found to behave as wave, and at other times, it appears to be corpuscular in character. For example, in the explanation of the phenomena like the photoelectric effect and the Compton effect, we have to assume that radiation is corpuscular in character, whereas in the explanation of some of the optical phenomena like interference and diffraction it is necessary to assume that it is wave like in nature. Thus, depending on the need of the situation, we have to invoke the corpuscular or wavelike nature and even though a behaviour such as this is in complete contradiction to all physical experience, today scientists are reconciled to this Jekyl and Hyde nature of radiation.

Confusion☹

> *J.J. Thompson wrote that the struggle between the two models was like "a struggle between a tiger and a shark: each is supreme in his own element, but helpless in that of the other"*
>
> *Banesh Hoffmann has written on the impact of this puzzle on Physicists of the time in his delightful and hilarious book "The strange story of quantum". He wrote they could but make the best of it, and went around with woebegone faces, sadly complaining that on Mondays, Wednesdays and Fridays they must look light as wave, on Tuesdays, Thursdays and Saturdays as a particle. On Sundays, they simply prayed.*

To add to this dilemma, Louis de Broglie proposed in 1924 that this duality should apply to matter also, thus leading to the concept of *'matter waves'*. We know that nature manifests itself in two fundamental forms, namely, matter and radiation. de Broglie reasoned that, since nature loves symmetry, if one form of nature is exhibiting duality in properties, the other form also must be equally capable of exhibiting dualistic behaviour. Alternatively, in other words, every material particle must also have a wave associated with it. This brilliant prediction was equally brilliantly verified by Davisson and Germer of the Bell Telephone Industries, USA in 1927. Using a nickel crystal as a diffraction grating, they were able to obtain diffraction patterns with a beam of electrons, thereby proving that electrons also have wave nature. At the same time, similar diffraction effects were also observed by Thompson by using extremely thin films of metal as diffraction gratings. Further confirmation of the idea of the association of waves and matter was provided by the diffraction phenomenon observed with the particles of hydrogen and helium. Theoretically, such effects should exist for all particles, but when the masses are relatively high, the equivalent wavelengths are too small to be measured.

We have already seen that $E = h\nu$ and the frequency is a variable that is associated with wave motion, whereas the energy of the system can be expressed in terms of particle concepts such as mass and velocity.

According to the theory of relativity the energy of a particle of mass 'm' and velocity 'c' is given by $E = mc^2$.

By equating the above two $h\nu = mc^2$

or $\quad h\nu/c = mc = p$

where 'p' is the momentum of the particle

or $\quad h/\lambda = mc = p$

If we now consider a material particle of mass 'm' and velocity 'v' the wavelength is given by $\lambda = h/mv = h/p$, such a wavelength is often referred to as de Broglie's wavelength.

de Broglie's hypothesis

> There is associated with the motion of each material particle a "fictitious wave" that somehow guides the motion of its quantum of energy. Using the methods of classical optics to describe the propagation of quanta, de Broglie was able to explain how photon (for that matter electron) diffract and interfere. It is not the particles themselves, but rather their "guide waves" that diffract and interfere.

Wave length and momentum of a particle

The photon is a relativistic particle of rest mass $m_o = 0$.

Therefore, $P = E/C$ ($m_o = 0$ and E = Total energy)

But $E = h\nu \qquad \therefore P = \dfrac{h\nu}{c} = \dfrac{h}{\lambda}$

Chapter 2 | The Wave Equation

In contrast to a photon, a material particle such as an electron has a non-zero rest mass, m_o. Therefore, the relationship between the energy and momentum of a particle moving at relativistic velocities (in a region of zero potential energy) is

$$E^2 = P^2C^2 + m_0^2C^4$$

Therefore Kinetic Energy, $T = E - m_0C^2$ or if $v \ll C$, $T = \dfrac{P^2}{2m_0}$. In either case the derivation of $P = \dfrac{h}{\lambda}$ cannot be applied to material particle.

However, de Broglie proposed that $P = \dfrac{h}{\lambda}$ and $E = h\nu$ can be used for material particles and photons. Notice that the de Broglie equation $\lambda = \dfrac{h}{P}$ implies an inverse relationship between the total energy E of a particle and its λ; namely

$$\lambda = \dfrac{hc/E}{\sqrt{1-\left(m_0c^2/E\right)^2}} \text{ for photon } \lambda = \dfrac{hc}{E} \text{ as } m_0 = 0.$$

de Broglie based his arguments on special theory of relativity. First he equated the rest energy, m_0c^2 of a material quantum to the energy, $h\nu_0$ to its 'periodic internal motion' where 'ν_0' is the intrinsic frequency of the particle. Next he considered a quantum moving at a velocity 'v' with momentum,

$$P = \dfrac{m_0 v}{\sqrt{1-\dfrac{v^2}{c^2}}}$$

and used relativistic kinematics to show that the frequency and wavelength of such a particle are given by the above equations.

Strange coincidence	It has been pointed out by M. Jammer that J. J. Thompson in 1906 (discharges through gases) was awarded the Nobel Prize for showing that the electron is a particle, and his son, G. P. Thompson in 1937 (electron diffraction of crystals) was awarded the Nobel Prize for showing that the electron is a wave.

2.2 Heisenberg's Uncertainty Principle

In the wave nature of the electron, we find the first of the two underlying precepts of quantum mechanics. The second of these is the Heisenberg's uncertainty principle, which states that it is impossible to determine the position and momentum of any particle precisely and simultaneously. The statement, which was first enunciated by Heisenberg in 1927, can be illustrated by the following discussion.

Suppose we device a hypothetical experiment to measure the position and velocity of an electron. We would set up two 'γ' ray microscopes that could see electrons, and measure the time taken for the electron to pass from one to the other.

P in the Fig.2.2.1 represents the electron. It can only be observed, if a photon incident upon it is scattered into the aperture of the microscope, i.e. within the cone of the angle 2α. Now a photon of frequency 'ν' will have an associated wavelength λ and hence λ = h/p = c/ν.

Therefore, p = hν/c.

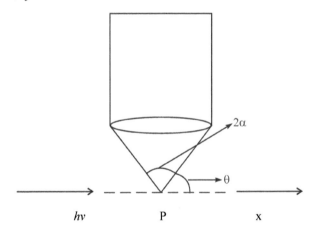

Fig. 2.2.1 The 'γ' ray microscope.

If the photon is scattered in a direction making an angle 'θ' with the 'x' axis, the electron will receive a component of momentum

$$\frac{h\nu}{c}(1-\cos\theta)$$

along the 'x'-axis and since the electron will be detected for any value of the angle 'θ' between $90^0 \pm \alpha$, the momentum may have any value between

$$\frac{h\nu}{c}\left[1-\cos(90-\alpha)\right] \text{ and } \frac{h\nu}{c}\left[1-\cos(90+\alpha)\right]$$

i.e. between $\frac{h\nu}{c}[1-\sin\alpha]$ and $\frac{h\nu}{c}[1+\sin\alpha]$.

If we define this spread of values, as the 'uncertainty' in the value of 'p' and denote it as 'Δp', then

$$\Delta p = \frac{2h\nu}{c}\sin\alpha$$

we would try to reduce this uncertainty by making 'α' small, i.e. by using a microscope of smaller aperture; but the accuracy with which an object can be located by a microscope is defined by the Rayleigh equation for the resolving power,

$$\Delta x = \frac{c}{\nu \sin\alpha}$$

where Δx is the uncertainty in 'x', the coordinate denoting the position of the electron.

Thus a smaller aperture, while decreasing the uncertainty in momentum, would increase the uncertainty in position.

In this experiment

$$\Delta x.\Delta p = \frac{2h\nu}{c}\sin\alpha . \frac{c}{2\sin\alpha} = 2h.$$

In general, the product, $\Delta x.\Delta p$ is of the order of the Planck's constant, 'h'. This is one way of expressing Heisenberg's uncertainty principle (1927).

We can illustrate its importance by an example. Suppose that we can locate the position of an electron with an uncertainty of 0.001Å, i.e. $\Delta x = 10^{-11}$ cm. We know that $\Delta x.\Delta p = h$

Substituting in equation

$$\Delta p = 6.6 \frac{10^{-27} g.cm^2.\sec^{-2}}{10^{-11} cm.\sec^{-1}} = 13.2 \times 10^{-16} g.cm.\sec^{-1}$$

This uncertainty in momentum, which is the result of uncertainty in velocity, is quite negligible in macroscopic systems, but it is far from negligible in systems containing electrons, since, there we are dealing with masses of the order of 10^{-27} grams. So precise statements of the position and momentum of the electrons have to be replaced by statements of probability that the electron has a given position and momentum.

The introduction of probabilities into description of electronic behaviour is a direct consequence of the uncertainty principle; a small uncertainty in position implies a high probability that the electron is at a given point.

This probability concept can be further illustrated, if we consider the electron diffraction experiments of Thompson in which the diffraction rings obtained correspond to regions of high electron density. If a single electron is sent through the diffraction apparatus, it obviously cannot interfere with itself to give a diffraction pattern, and the Heisenberg's uncertainty principle tells us that we cannot follow its course precisely. We can say, however, that there is a certain probability that it will take a particular path, and that the electron is most likely to be found in those regions where we get the greatest electron density in experiments using beams of electrons. Thus, a high intensity in a diffraction experiment, measured by the square of an amplitude factor in a wave equation, can be related to a high probability, that an electron is in a unit volume around a given point.

We will make use of both the concepts, electron density and electron probability.

It is not difficult to see why the uncertainty relation should exist. Any measurement must, by necessity, result in some disturbance on the system. Thus, when we determine the position of a quantum mechanical object, say, an electron, we have also supplied some energy to it (for example by shining light on it) so that its velocity or momentum becomes less well defined. However, in dealing with macroscopic bodies, the amount of perturbation is so negligibly small that its momentum can be accurately measured at the same time.

The Heisenberg's uncertainty principle can also be expressed in terms of energy and time as follows.

Since, momentum/time = force and energy = force × distance, we write

$$h/2\Pi = \Delta(\text{momentum}).\Delta(\text{distance})$$
$$= \Delta(\text{force x time}).\Delta(\text{distance})$$
$$= \Delta(\text{force x distance}).\Delta(\text{time})$$
$$= \Delta(\text{energy}).\Delta(\text{time}) = \Delta E.\Delta t$$

Thus, we cannot measure the kinetic energy of a particle with absolute precision (that is, to have $\Delta E = 0$) in a finite span of time. This equation is particularly useful for estimating spectral line widths.

The uncertainty principle comes into force because of the wave like properties of "particles" like electrons and protons. The waves tell the dynamics of the particle – its momentum, its energy and even its angular momentum.

The wave runs through time and space. The wavelength of the wave running through space gives the particle's momentum. The frequency of the wave running through time gives the particle's energy. *However, a wave cannot really represent a particle*. A particle is located in, at only one place in space. A wave is not located in only one place.

The conflict between a wave and a particle can never be resolved. It can only be compromised, and can only be compromised by the '*uncertainty principle*'. The compromise goes like this: If you get a wave that will not run forever, but will just wave, in one location, and then kill itself, it will be like a particle. This kind of wave is called a 'wave packet'.

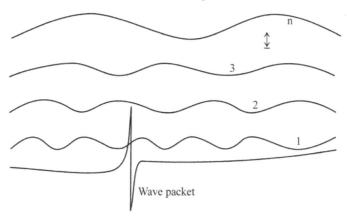

Fig. 2.2.2 A wave packet made of simple harmonic waves (1,2,3 ..n).

Wave packets are made by adding together many simple (harmonic) waves that run forever. They do this because the waves have different wavelengths and frequencies, so some are in phase while others are not. However, they do not cancel each other completely. *At one special place, the location where you want the particle to be, all the waves should be in phase. That way the added effect of the waves is constructive in that one special location and destructive everywhere else.*

Therefore, it takes a mixture of different waves to make a wave packet, which we call a particle as shown in Fig.2.2.2. Now comes the catch. The wavelength or frequency represents the momentum or energy of a particle. If a mixture of different waves makes a particle, the particle automatically has a mixture of momenta and energies. That mixture is the uncertainty. Of course, you can make the particle out of only one wave, so that there is no uncertainty in momentum and energy. But, one wave will not make a wave packet and it runs forever. So, if you make a particle from one wave, you cannot tell, at what location or what time it exists. Uncertainty again. However, you do not have to have uncertainty in momentum or energy - if you are willing to accept uncertainty in location or time – if you accept uncertainty in momentum or energy you can avoid uncertainty in location or time. *You can get some uncertainty in anything, but you cannot rid of all the uncertainty in everything.*

The number 'h' known as Planck's constant tells us how much uncertainty there must always be. The quantity 'h' is a basic constant of universe. The product of the two uncertainties together has a minimum value of 'h'. It does not become obvious until you enter into the world of photons and electrons.

Energy & Time and Position & Momentum are conjugate variables.

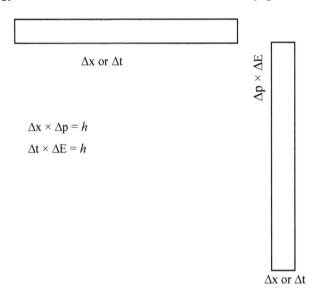

2.3 Wave Equation

If electrons have the wave properties then there must be a wave equation and a wave function to describe the electron waves just as the waves of light, sound and strings are described. Let us consider the motion of a string which is held fixed at two ends, $x = 0$ and $x = a$. It is possible to excite with care certain kinds of vibrations, in which all points of the string move, so that their displacements vary with time in the same way and all points are at their maximum displacements at the same time and have their maximum velocity at the same time. If the displacements occur in the y-direction, mathematically these motions can be described by functions of the form

$$y(x,t) = f(x)\Phi(t) \qquad \ldots(2.3.1)$$

where $f(x)$ is independent of t and $\Phi(t)$ is independent of x. Such motions are called normal modes of vibration. The wave equation has the general form

$$\frac{d^2 y}{dx^2} = \frac{1}{c^2}\frac{d^2 y}{dt^2} \qquad \ldots(2.3.2)$$

where 'c' is called the wave velocity. Substituting for 'y' from the Eq. 2.3.1 in Eq.2.3.2 one obtains

$$\frac{c^2}{f(x)}\frac{d^2f(x)}{dx^2} = \frac{1}{\Phi(t)}\frac{d^2\Phi(t)}{dt^2} = -\omega^2 \qquad .....(2.3.3)$$

In Eq. 2.3.3 the variables are separated and they may be equated to the same constant, say $-\omega^2$. This gives us two ordinary differential equations

$$\frac{d^2\Phi(t)}{dt^2} + \omega^2\Phi(t) = 0 \qquad .....(2.3.4)$$

$$\frac{d^2f(x)}{dx^2} + \omega^2\frac{f(x)}{c^2} = 0 \qquad .....(2.3.5)$$

Eq.2.3.4 has the solution

$$\Phi(t) = A\sin\omega t + B\cos\omega t \qquad .....(2.3.6)$$

where the two constants A and B are determined from the laboratory conditions, and ω is called the circular frequency which is related to the ordinary frequency 'v' as

$$\omega = 2\pi v \qquad .....(2.3.7)$$

Eq. 2.3.5 may therefore be written as

$$\frac{d^2f(x)}{dx^2} + \frac{4\pi^2 v^2}{c^2}f(x) = 0 \qquad .....(2.3.8)$$

Setting $\lambda = c/v$ the general solution of Eq. 2.3.8 may be written as

$$\frac{d^2f(x)}{dx^2} + \frac{4\pi^2}{\lambda^2}f(x) = 0$$

$$f(x) = A_1 \exp\left(+\frac{i2\pi x}{\lambda}\right) + A_2 \exp\left(\frac{-i2\pi x}{\lambda}\right) \qquad .....(2.3.9)$$

or

$$f(x) = c\sin\left(\frac{2\pi}{\lambda}x\right) + D\cos\left(\frac{2\pi}{\lambda}x\right) \qquad .....(2.3.10)$$

where A_1, A_2, C and D are constants. Let us consider Eq. (2.3.10) and impose the boundary conditions

(i) $f(x) = 0$ at $x = 0$; and
(ii) $f(x) = 0$ at $x = a$

where 'a' is the length of the string. From the boundary condition (i), D = 0 and from the condition (ii), $c\sin\frac{2\pi}{\lambda}a = 0$ or $\sin\frac{2\pi}{\lambda}a = 0$ or

$$\frac{2}{\lambda}a = n, \text{ where } n = 1,2,3,\ldots\ldots \quad \ldots(2.3.11)$$

where n is a positive integer. Thus

$$a = n\frac{\lambda}{2} \quad \ldots(2.3.12)$$

The normal modes are thus the stationary sine waves given by

$$f(x) = c\sin\frac{n\pi}{a}x \quad \ldots(2.3.13)$$

and the wavelengths λ are such that the length of the string is an integral number of half waves. The complete solution for a normal mode in a stretched string therefore follows from Eqs. 2.3.6, 2.3.7, 2.3.11 and 2.3.13 and is given by

$$y(x,t) = C\sin\frac{n\pi}{a}x \cdot (A\sin 2\pi vt + B\cos 2\pi vt) \quad \ldots(2.3.14)$$

Eq. 2.3.14 is an expression for the amplitude of waves generated during the normal modes of vibration in a stretched string. The same equation should represent the amplitude of a de Broglie wave associated with a moving particle. We are primarily concerned here, with the time independent or stationary waves. Therefore, the equation for a standing sine wave of wavelength λ is given by

$$\psi = C\sin\frac{n\pi}{a}x = C\sin\frac{2\pi}{\lambda}x \quad \ldots(2.3.15)$$

where 'ψ' is called the wave function with the amplitude of the wave varying sinusoidally along x and C is the maximum amplitude. Double differentiation of Eq. 2.3.15 with respect to x gives

$$\frac{d^2\psi}{dx^2} = -\frac{4\pi^2}{\lambda^2}C\sin\frac{2\pi}{\lambda}x = -\frac{4\pi^2}{\lambda^2}\psi \quad \ldots(2.3.16)$$

The kinetic energy T of a moving particle of mass m and velocity 'v' is given by

$$T = \frac{1}{2}mv^2 = \frac{m^2v^2}{2m} \quad \ldots(2.3.17)$$

Following the de Broglie relation, T becomes

$$T = \frac{h^2}{2m\lambda^2} \quad \ldots(2.3.18)$$

By using Eq. 2.3.16 to eliminate λ^2 from Eq. 2.3.18 we get

$$T = -\frac{h^2}{8\pi^2 m} \cdot \frac{1}{\psi}\frac{d^2\psi}{dx^2} \quad \ldots(2.3.19)$$

If the particle moves in a field, whose potential energy is V, then

$$T = E - V = -\frac{h^2}{8\pi^2 m} \cdot \frac{1}{\psi} \frac{d^2\psi}{dx^2} \quad \ldots(2.3.20)$$

where E is the total energy. This is Schrödinger's equation for a particle in one dimension. It is usually written as

$$\frac{d^2\psi}{dx^2} + \frac{8\pi^2 m}{h^2}(E - V)\psi = 0 \quad \ldots(2.3.21)$$

In three dimensions this equation becomes:

$$\frac{d^2\psi}{dx^2} + \frac{d^2\psi}{dy^2} + \frac{d^2\psi}{dz^2} + \frac{8\pi^2 m}{h^2}(E - V)\psi = 0 \quad \ldots(2.3.22)$$

2.4 Interpretation of Wave Function

The success of wave mechanics was well demonstrated by Erwin Schrödinger before an acceptable interpretation of the wave function was known. Max Born utilized the probability concepts of the uncertainty principle to give us the presently accepted ideas of the wave function. According to Born, the wave function of a particle is not an amplitude function in the common sense used for the ordinary waves, but rather, it is a measure of the probability of a mechanical event.

It might then be expected that a quantum interpretation that seems quite reasonable for photon should also hold for an electron. This leads to the postulate that the square of the wave function of an electron is proportional to the probability of finding the electron in a given volume element dx.dy.dz. Such an interpretation is just a postulate and may or may not be legitimate. One of the most significant indications of its validity lies in the treatment of directional bonding in molecules. The positions, at which the density of the bonding electrons is calculated to be the greatest, are where the bonded atoms are found to be located. For example, in the molecule H_2S, the hydrogen atoms lie at an angle of about 92^0 with respect to each other, and according to simple theoretical calculations the electron density is a maximum at an angle of 90^0.

The symbol ψ is usually used to denote the wave function of an electron, and very often contains, 'i' the square root of '–1'. Since the probability that an electron is in a given volume element must be a real quantity, the product $\psi\psi^*$ will always be real, where ψ can possibly be imaginary. As an example $a + ib$ can be considered to be the complex quantity. Its complex conjugate can be obtained by changing 'i' to '$-i$' giving $a - ib$. The product will then be $a^2 + b^2$, which is always a real quantity.

Ex. If $\psi = a + ib$

$\psi^* = a - ib$

The product $\psi\psi^* = (a+ib)(a-ib) = a^2 + b^2$

If ψ turns out to be a real quantity initially, then ψ and its complex conjugate are the same.

2.5 Normalized and Orthogonal Wave Functions

The square of the wave function is said to be proportional rather than equal to the probability that the electron is in a given volume element *dxdydz*. This arises from the fact that if the wave function, ψ is a solution to the wave equation, multiplication by any constant such as A will give a new wave function, A ψ, which is also a solution to the wave equation. This means that it is not possible to say that ψ^2 is equal to the probability, but it is only proportional to the probability that the electron is in the given volume element. However, since multiplication by a constant is possible, it is usually convenient to multiply the wave function by a constant that will make the square of the resultant wave function equal to probability.

The probability of a certainty is defined as unity. Thus, if it is a known fact that the electron is in a given volume element, *dxdydz*, then we can say that the probability that it is in this volume element is unity. This leads to the relation

$$\int \psi\psi^* dxdydz = 1$$

If a wave function satisfies this relation, it is said to be normalized. If the electron is in the volume element, *dxdydz* then $\int \psi\psi^* dxdydz$, will be equal to the probability that the electron is in this volume element.

Very often ψ is not a normalized wave function. However, we know that it is possible to multiply ψ by a constant, A, to give a new wave function A ψ, which is also a solution to the wave equation. The problem is to choose the proper value of A to make the new wave function A ψ normalized function. In order for the new wave function, A ψ, to be a normalized function, it must meet the requirement

$$\int A\psi A\psi^* dxdydz = 1$$

Since A is a constant, it can be removed from under the integral sign giving

$$A^2 \int \psi\psi^* dxdydz = 1$$

A is known as a normalizing constant and can be determined from the above equation.

If ψ_1 and ψ_2 are two wave functions then these two wave functions are orthogonal to each other in case

$$\int \psi_1 \psi_2 \, dxdydz = 0$$

If ψ_1 and ψ_2 are also normalized wave functions then they are called orthonormal wave functions.

2.6 Exercises

Q.1: *Determine the value of A to make the wave function,*

$$\psi_N = A \sin \frac{n\pi x}{a}$$

a normalized wave function within the limits x = 0 to a.

Solution:

Let A is the normalization constant and ψ_N the normalized wave function.

Then

$$\psi_N = A \sin \frac{n\pi x}{a}$$

within the limits $0 \to a$

The normalizing condition is

$$\int_0^a \psi_N \psi_N^* \, dx = 1$$

$\therefore$ Substituting the value of ψ_N in the above equation

$$\int_0^a A \sin \frac{n\pi x}{a} A \sin \frac{n\pi x}{a} \, dx = 1$$

$$\int_0^a \left(A \sin \frac{n\pi x}{a} \right)^2 dx = 1$$

$$\int_0^a A \sin^2 \left(\frac{n\pi x}{a} \right) dx = \frac{1}{A^2} \quad \text{but} \quad \sin^2 \theta = \frac{1 - \cos^2 \theta}{2}$$

Substituting

$$\int_0^a \frac{1-\cos(2n\pi x/a)}{2}dx = \frac{1}{A^2}$$

$$\int_0^a \frac{1}{2}dx - \int_0^a \frac{\cos(2n\pi x/a)}{2}dx = \frac{1}{A^2}$$

$$\frac{1}{2}[x]_0^a - \frac{1}{2}\left[\frac{\sin(2n\pi x/a)}{2n\pi/a}\right]_0^a = \frac{1}{A^2}$$

The normalized wave function is

$$\psi_N = \sqrt{\frac{2}{a}}\sin\frac{n\pi x}{a}$$

CHAPTER 3

The Postulates

3.1 The Formulation of Quantum Mechanics

In its beginning, quantum mechanics was approached in two completely different ways. Schrödinger, reasoning that electronic motions could be treated as waves, developed wave mechanics. In this treatment, he took over the great body of information from classical physics about wave motion and applied it to electronic and molecular motions. The stationary states that an electron or molecule might have were analogues to standing waves set up by applying appropriate boundary conditions. Later on, a mathematical formalism becomes associated with the Schrödinger method that related observable quantities to certain mathematical operators. Werner Heisenberg, independently and slightly earlier, had used the properties of matrices to get the same results as Schrödinger. This approach to quantum mechanics looked very different, but a little later Born and Jordan showed that they are equivalent. Later still, in the more general treatments of quantum mechanics by P. A. M. Dirac and J. Von Neumann, the Schrödinger and Heisenberg approaches were shown to be specific cases of a more general theory.

Schrödinger Wave Equation

Bohr's theory could not explain the Stark and Zeeman effect, and the spectra of atoms more complex than alkali metals. With the formulation of wave particle duality of matter by de Broglie in 1924, it was possible to treat particles such as electrons as waves and a wave equation for such a purpose was sought. de Broglie related the wavelength of a wave associated with the linear momentum, P = mv, of a particle by the celebrated equation $\lambda = \dfrac{h}{p}$ or with the energy

$$\lambda = \dfrac{h}{\sqrt{2m(E-V)}}.$$

In 1927, Heisenberg added one more dimension to the problem through the uncertainty principle. The uncertainty principle states that "it is impossible to determine precisely and simultaneously both the position and momentum of an electron". So the product of the uncertainties in linear momentum and in position is at least of the order of Planck's constant, i.e.., $\Delta x.\Delta p \approx h$,

Both these important relations are incorporated in the wave equation constructed by Schrödinger and known after his name. It is written as,

$$\nabla^2 \psi + \frac{8\pi^2 m}{h^2}(E-V)\psi = 0$$

where, $\nabla^2 = \dfrac{\partial^2}{\partial x^2} + \dfrac{\partial^2}{\partial y^2} + \dfrac{\partial^2}{\partial x^2}$

E = Total Energy, V = Potential Energy,

E-V = T = Kinetic Energy,

m = mass,

ψ = the wave function which takes care of the wave nature, of the particle having mass 'm'.

Such a wave equation can easily be derived based on postulates of quantum mechanics.

3.2 The Postulates of Quantum Mechanics

The postulates of any theory are a set of fundamental statements that are asked to be believed and draw conclusions from them. These conclusions are then tested by experiment and if they are confirmed, the belief in postulate is justified. Before going into the postulates themselves, it is necessary to understand the meaning of the terms "dynamical variables" and "observables".

Any property of a system of interest is called a dynamical variable. Thus the position 'r', the energy 'E', the 'x' component of the linear momentum P_x, and so on, are dynamical variables even though in a given system some of them may be constant. In general, any quantity of interest in classical mechanics is a dynamical variable. An observable is any dynamical variable that can be measured. In classical mechanics, all dynamical variables are observables, but there are certain fundamental restrictions placed upon simultaneously measurable quantities in quantum mechanics. To measure the components of the momentum vector, which the particle has at some point P, it is necessary to make a simultaneous measurement of the position and momentum of the particle. However, there exists an uncertainty relation for such a simultaneous measurement of dynamical variables on microscopic particles. With this background in mind, we introduce the basic postulates of quantum theory.

3.2.1 Postulate I

(a) Any state of dynamical system of 'N' particles is described as fully as possible by a function, $\psi(q_1, q_2, q_3, \ldots q_{3n}, t)$ such that

(b) the quantity $\psi\psi^* d\tau$ is proportional to the probability of finding q_1 between q_1 and $q_1 + d_{q1}$; q_2 between q_2 and $q_2 + d_{q2}$;............, q_{3n} between q_{3n} and $q_{3n} + d_{q3n}$ at a specific time 't'.

What this postulate says is that all the information about the properties of a system is contained in a 'ψ' function which is a function only of the coordinates of the 'N' particles and the time 't'. If the wave function includes the time explicitly, it is called the time dependent wave function. If the observable properties of a system do not change with time, the system is said to be in a stationary state. A 'ψ' function describing such a state is called a stationary state wave function, and the time dependence of such a wave function can be separated out.

The second part of the postulate gives a physical interpretation of the 'ψ' function. This interpretation is the easiest to visualize for a system containing a single particle constrained to move in one dimension. The quantity dx is then the probability of finding the particle between x and $x + dx$ at a given time 't'. A 'ψ' function may be complex; hence the probability density is a product of ψ with its complex conjugate.

In order for these functions to be in accord with physical reality, they are subject to certain restrictions. These restrictions are the following:

3.2.2 Well Behaved Wave Function

Even in the weird world of quantum mechanics, the condition that a particle must exist somewhere, restricts the class of physically admissible functions to those that are normalizable. But, normalizability is the only mathematical condition a function should satisfy, if it is to represent a quantum state.

1. The first of these restrictions follows from the Born interpretation of the state function, as a position probability amplitude. At any time the value of the probability of finding the particle in an infinitesimal region of space must be unique. This implies that the state function must assume only one value at each time i.e., it must be '*Single Valued*'.

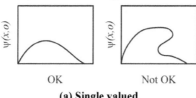

OK   Not OK

(a) Single valued

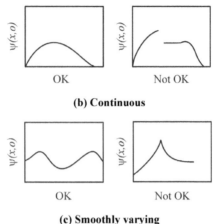

(b) Continuous

(c) Smoothly varying

Additional restrictions follow from the equation of motion of quantum mechanics.

Consider the first term, the second derivative with respect to x. We cannot even define this derivative unless the first derivative of the function in question is continuous. Mathematicians call a function, whose first derivative is everywhere continuous, a smoothly varying function. Therefore, the wave function should be continuously varying and should not have any kinks.

Another condition also follows from the presence of second derivative in the Schrödinger equation. We define the second derivative in terms of the first derivative, and we can define the first derivative, only if, the function being differentiated, is itself continuous. Incidentally, the Born interpretation provides another incentive to require that ' ψ ' be continuous: if it weren't, then at a point of discontinuity, the value of its modulus squared would not be unique, in which case we cannot meaningfully interpret this value as the position probability density.

The fitness of the wave function

State functions are usually finite at all points. However, do not misconstrue this observation as a requirement. A state function can be finite at a point only if the mathematical nature of this singularity is such that the normalization integral formed from the function is finite.

- The function should be continuous and smooth. This implies that its first derivative will be continuous as well. If the first derivative is not continuous, the second derivative cannot be defined.
- The function should be single valued.

- The function should have an integrable square. This requirement can be interpreted, to mean that the function is everywhere finite. It also means that 'ψ' will go to zero at $\pm \infty$.

These restrictions all arise from the postulate that $\psi\psi^* d\tau$ represents a probability. The restriction of integrable squares is simply the requirement that the probability of finding the system in all space must be finite. A special case of this requirement is when the integral

$$\int_{-\infty}^{+\infty} \psi\psi^* d\tau = 1$$

When this is true, the function is said to be normalized. The physical meaning of this, for a single particle system is that, the probability of finding the particle in some region in space must be '1' i.e., the probability of a certainty is defined as unity.

3.2.3 Postulate II

For every observable property of a system, there exists a linear Hermitian Operator.

Let us first define the four new terms found here.

1. **Observables:** These are properties of a system that can be experimentally determined. Thus position, x; velocity, v; linear momentum, p; angular momentum, L; potential energy, V; kinetic energy, T; total energy, E are some observables.
2. **Operator:** An operator is a mathematical symbol, which tells us to carry out an operation. It is represented by a tent or circumflex on its sign Eg. $\hat{O}$. Thus in the expression $\sqrt{2}$, the $\sqrt{}$ is an operator telling one to take the square root of what follows, in this case 2.
3. **Linear operator:** In quantum mechanics, luckily the operators used are of limited type. They are linear only. A linear operator obeys the following rule.

$$\hat{O}(ax^2 + bx + c) = \hat{O}(ax^2) + \hat{O}(bx) + \hat{O}(c)$$

Thus, a linear operator operates on each quantity within the bracket separately. E.g. Operator of integration, differentiation etc.

Here is an operator telling one to take the derivative with respect to 'x' of what follows, that is $\frac{d}{dx}$. The algebra of operators follows definite mathematical procedures.

Thus if, $\hat{P} = \left(\dfrac{\partial}{\partial x}\right)_{yz}$ and $\hat{Q} = \left(\dfrac{\partial}{\partial y}\right)_{xz}$

$$\hat{P}\hat{Q} = \left[\dfrac{\partial}{\partial x}\left(\dfrac{\partial}{\partial y}\right)_{xz}\right]_{yz} = \dfrac{\partial^2}{\partial x \partial y}$$

The operator given above, it turns out that $\hat{P}$ and $\hat{Q}$ do commute.

That is, $\hat{P}\hat{Q} = \hat{Q}\hat{P}$

Since $\dfrac{\partial^2}{\partial x \partial y} = \dfrac{\partial^2}{\partial y \partial x}$

but, this in general will not be the case.

The quantity $\hat{P}\hat{Q} - \hat{Q}\hat{P}$ is the commutator of $\hat{P}$ and $\hat{Q}$. It is often symbolized as $\left[\hat{P},\hat{Q}\right]$. If $\hat{P}$ and $\hat{Q}$ commute, then the value of the commutator is zero. Conversely, if the value of the commutator is zero, the operators $\hat{P}$ and $\hat{Q}$ commute.

If an operator $\hat{P}$ is complex, the complex conjugate, $\hat{P}^*$ is formed by replacing 'i' by '-i', wherever it occurs.

Thus, if $\hat{P} = \dfrac{id}{dx}$ and $\hat{P}^* = -\dfrac{id}{dx}$

In quantum mechanics, only linear operators are used. An operator is linear if it is true that,

$$\hat{P}(f+g) = \hat{P}f + \hat{P}g$$

and $\hat{P}af = a\hat{P}f$

where 'a' is a constant. One may easily verify that $\dfrac{d}{dx}$ is a linear operator, whereas $\sqrt{\ }$ is not.

Hermitian operator

A Hermitian operator makes the calculation of an observable real. A Hermitian operator is defined by the relation

$$\int_{-\infty}^{+\infty} \psi_i^* \hat{\alpha} \psi_j d\tau = \int_{-\infty}^{+\infty} \psi_j \left(\hat{\alpha} \psi_i \right)^* d\tau$$

where ψ_i^* and ψ_j are any two wave functions which satisfy the conditions for acceptability stated above and $\hat{\alpha}$ is an operator of interest.

At this point, it is convenient to introduce a new notation for integrals of the type used above. This notation represents the integration over all space by parentheses or angular bracket.

Thus,

$$\int_{-\infty}^{+\infty} \psi_i^* \hat{\alpha} \psi_j d\tau = \langle \psi_i | \hat{\alpha} | \psi_j \rangle$$

$$\int_{-\infty}^{+\infty} \psi_i^* \psi_j d\tau = \langle \psi_i | \psi_j \rangle$$

$$\langle \hat{\alpha} \rangle = \frac{\langle \psi_s | \hat{\alpha} | \psi_s \rangle}{\langle \psi_s | \psi_s \rangle} = \frac{\int_{-\infty}^{+\infty} \psi_s^* \hat{\alpha} \psi_s d\tau}{\int_{-\infty}^{+\infty} \psi_s^* \psi_s d\tau}$$

and $$\int_{-\infty}^{+\infty} \psi_i^* \psi_j d\tau = \langle \psi_i | \psi_j \rangle$$

The condition for Hermitian operator in this notation becomes

$$\langle \psi_i | \hat{\alpha} | \psi_j \rangle = \langle \psi_j | \hat{\alpha} | \psi_i \rangle^*$$

Coming back to postulate II, we find that every observable has its linear Hermitian operator. Since we are interested in observables, which are real, the operator, $\hat{O}$ has to be Hermitian. The question naturally arises, as to how; one gets the operators for a given observable. First, the classical expression for the observable of interest is written down in terms of coordinates, momenta and time. Then classical expression for the operator is changed into the quantum mechanical form of the operator. This can be done as per the following.

These operators have been derived for many observables.

1. The quantum mechanical form of the operator for Cartesian Coordinates x, y and z are the same as those for the classical representation.

2. The potential energy 'V' and time 't' are unchanged in both the systems. The linear momentum, $P_x = mv_x$ in classical mechanics, is replaced by the quantum mechanical operator,

$$\hat{P}_x = \frac{-ih}{2\pi}\left(\frac{\partial}{\partial x}\right)$$

for a single particle moving along x-axis.
Similarly for y and z

$$\hat{P}_y = \frac{-ih}{2\pi} \text{ and } \hat{P}_z = \frac{-ih}{2\pi}\left(\frac{\partial}{\partial z}\right).$$

Similarly for angular momentum, the classical form is replaced by the quantum mechanical operator.

$$L_x = yP_z - zP_y = \frac{-ih}{2\pi}\left(y\frac{\partial}{\partial z} - z\frac{\partial}{\partial y}\right)$$

$$\text{and } L_y = zP_x - xP_z = \frac{-ih}{2\pi}\left(z\frac{\partial}{\partial x} - x\frac{\partial}{\partial z}\right)$$

$$L_z = xP_y - yP_x = \frac{-ih}{2\pi}\left(x\frac{\partial}{\partial y} - x\frac{\partial}{\partial x}\right)$$

The operator for kinetic energy, T can be constructed from the relations from the classical mechanics namely,

$$T_x = \frac{1}{2}mv_x^2 = \frac{p_x^2}{2m}$$

The classical expression for the kinetic energy of a particle in Cartesian coordinates is,

$$T = \frac{1}{2m}\left(p_x^2 + p_y^2 + p_z^2\right)$$

Let us now construct the quantum mechanical operator for kinetic Energy, $\hat{T}$.

The momentum operator $\hat{P}_x = \frac{-ih}{2\pi}\left(\frac{\partial}{\partial x}\right)$

Substituting this,

$$\hat{T} = \frac{1}{2m}\left[\left(\frac{-ih}{2\pi}\frac{\partial}{\partial x}\right)^2 + \left(\frac{-ih}{2\pi}\frac{\partial}{\partial y}\right)^2 + \left(\frac{-ih}{2\pi}\frac{\partial}{\partial z}\right)^2\right]$$

$$\hat{T} = \frac{h^2}{8\pi^2 m}\left(\frac{\partial^2}{\partial x^2} + \frac{\partial^2}{\partial y^2} + \frac{\partial^2}{\partial z^2}\right)$$

Hence, the kinetic energy operator is,

$$\hat{T} = \frac{h^2}{8\pi^2 m}\nabla^2$$

Perhaps the most important operator that will concern us is the operator connected with the total energy, $\hat{E}$ of a system. The classical expression for the total energy is Hamiltonian function, and the corresponding operator is called the Hamiltonian. The expression for the Hamiltonian for a single particle system is,

$$\hat{H} = \hat{T} + \hat{V}$$

$\hat{T}$ has already been derived and $\hat{V}$ is only a function of coordinates, q, that according to our prescription, remain the same.

Therefore,

$$\hat{H} = -\frac{h^2}{8\pi^2 m}\nabla^2 + V(q)$$

3.2.4 Postulate III

Eigenfunctions and Eigenvalues: When an operator $\hat{\alpha}$ operates on a wave function, ψ, such that the resultant is a constant times, ψ. That is,

$$\hat{\alpha}\psi = \text{Constant} \times$$
$$\hat{\alpha}\psi = m x$$

where 'm' is a number.

Such a wave function is called the eigenfunction of the operator and the constant 'm' is called the eigenvalue of the same operator.

This is one of the postulates that bridges the gap between the mathematical formalism of quantum mechanics, and the experimental measurements in the laboratory.

> *Experiments are the only means of knowledge at our disposal. The rest is poetry and imagination.*   Max Planck

The above equation is basic to a large number of calculations in quantum mechanics as this gives us a method of obtaining the observables. This, if we operate upon a wave function with the operator for linear momentum, $\hat{P}_x$; if the result is the number times the same wave function, then this number (constant) is the linear momentum P_x,

$$\hat{P}_x \psi = P_x \psi$$

Suppose one is interested in calculating the allowed energies in a molecular or atomic system, one has to use the Hamiltonian operator, $\hat{H}$ the total energy operator. Let the state of the system is described by a function, ψ, which is an eigenfunction of the operator corresponding to the total energy, the Hamiltonian.

Then the eigenvalue equation for this is,

$$\hat{H}\psi = E\psi$$

But

$$\hat{H} = -\frac{h^2}{8\pi^2 m}\nabla^2 + V(q)$$

Substituting this,

$$\left(-\frac{h^2}{8\pi^2 m}\nabla^2 + V\right)\psi = E\psi$$

Rearranging this,

$$\left(-\frac{h^2}{8\pi^2 m}\nabla^2 \psi\right) = (E - V)\psi$$

or

$$\nabla^2 = -\frac{8\pi^2 m}{h^2} + (E - V)\psi$$

i.e.

$$\nabla^2 \psi + \frac{8\pi^2 m}{h^2}(E - V)\psi = 0$$

This is the Schrödinger wave equation for a single particle in a stationary state.

If one is interested in calculating other properties of the system, such as the value of the angular momentum about the 'Z' axis, the procedure is the same, but the appropriate operator must be used in deriving the eigenvalue equation.

3.2.5 Postulate IV

Given an operator $\hat{\alpha}$ and a set of identical systems characterized by a function, ψ that is not an eigenfunction of a series of measurements of the property corresponding to $\hat{\alpha}$ on different members of the set will not give the same result. Rather, a distribution of results will be obtained, the average of which will be,

$$\langle \hat{\alpha} \rangle = \frac{\langle \psi_s | \hat{\alpha} | \psi_s \rangle}{\langle \psi_s | \psi_s \rangle} = \frac{\int_{-\infty}^{+\infty} \psi_s^* \hat{\alpha} \psi_s d\tau}{\int_{-\infty}^{+\infty} \psi_s^* \psi_s d\tau}$$

This is called the "mean value" theorem that tells what the experimental result will be when a system is not described by an eigenfunction of the operator involved. The symbol $\langle \hat{\alpha} \rangle$ is the average or expectation value of the quantity associated with $\hat{\alpha}$. The average value in quantum mechanics should not be confused with a time average in classical mechanics. Rather, it is the number average of a large number of measurements of the property corresponding to $\hat{\alpha}$. Obviously, if ψ_s is an eigenfunction of $\hat{\alpha}$, the average value will be the same as the eigenvalue.

Much modern research in quantum chemistry and spectroscopy is concerned with time dependent phenomena. In this case, the problem is to know how the state function $\psi(q,t)$ develops in time. We therefore introduce postulate V.

3.2.6 Postulate V

The evaluation of a state vector, $\psi(q,t)$ in time, is given by the relation,

$$\frac{ih}{2\pi} \frac{\partial \psi}{\partial t} = \hat{H} \psi$$

where $\hat{H}$ is the Hamiltonian operator for the system. The above equation is called the time dependent Schrödinger equation.

3.3 Exercises

1. Show that the eigenvalues of a Hermitian operator are real

Let $\hat{R}$ be a Hermitian operator with eigenvalue 'r'.
Then $\quad \hat{R}\psi = r\psi \quad$ (3.3.1)
Taking the complex conjugate of both sides, we have
$$\hat{R}^*\psi^* = r^*\psi^* \quad (3.3.2)$$
Multiplying Eq.3.3.1 by ψ^* and Eq.3.3.2 by ψ and integrating, yields

$$\int_{-\infty}^{+\infty} \psi^* \hat{R} \psi d\tau = \int_{-\infty}^{+\infty} \psi^* r \psi d\tau = r \int_{-\infty}^{+\infty} \psi^* \psi d\tau$$

and
$$\int_{-\infty}^{+\infty} \psi \hat{R}^* \psi^* d\tau = \int_{-\infty}^{+\infty} \psi r^* \psi^* d\tau = r^* \int_{-\infty}^{+\infty} \psi \psi^* d\tau$$

Since $\hat{R}$ is Hermitian

$$r \int_{-\infty}^{+\infty} \psi^* \psi d\tau = r^* \int_{-\infty}^{+\infty} \psi \psi^* d\tau$$

Therefore, $r = r^*$, i.e. eigenvalues of Hermitian operator are real.

2. Show that the eigenfunctions of any Hermitian operator are orthogonal

If $\hat{R}$ is a Hermitian operator then the corresponding eigenvalues for ψ_1 and ψ_2 may be determined as

$$\hat{R}\psi_1 = r_1 \psi_1 \qquad (3.3.3)$$

and $\qquad \hat{R}\psi_2 = r_2 \psi_2 \qquad (3.3.4)$

Multiplying (3.3.3) by ψ_2^* and integrating

$$\int \psi_2^* \hat{R}\psi_1 d\tau = r_1 \int \psi_2^* \psi_1 d\tau$$

Then applying the definition of a Hermitian operator and the fact that the eigenvalues be real we have

$$\int \psi_2^* \hat{R}\psi_1 d\tau = \int \psi_1 \hat{R}^* \psi_2^* d\tau = \int \psi_1 r_2^* \psi_2^* d\tau$$
$$= r_2^* \int \psi_1 \psi_2^* d\tau = r_2 \int \psi_1 \psi_2^* d\tau$$

That is

$$r_1 \int \psi_2^* \psi_1 d\tau = r_2 \int \psi_1 \psi_2^* d\tau \text{ or } (r_1 - r_2) \int \psi_2^* \psi_1 d\tau = 0$$

If $\qquad r_1 \neq r_2$ then $\int \psi_2^* \psi_1 d\tau = 0$.

Hence, the eigenfunctions of the Hermitian operator are real.

3. Calculate the linear momentum in 'X' direction for $\psi = Ae^{\pi x}$

Find out whether 'ψ' is an eigenfunction for the momentum operator, $\hat{P}_x$ or square of the momentum operator, $\hat{P}_x^2$.

If the operator operates on 'ψ' and we get the constant times the same wave function, then the wave function is the eigenfunction of the operator.

$$\hat{P}_x = -\frac{ih}{2\pi}\frac{d}{dx} \text{ and } \psi = Ae^{\pi x}$$

$$\hat{P}_x \psi = -\frac{ih}{2\pi}\frac{d}{dx}\psi = -\frac{ih}{2\pi}\frac{d}{dx}Ae^{\pi x} = -\frac{ih}{2}\left(Ae^{\pi x}\right) = -\frac{ih}{2}$$

Hence, 'ψ' is an eigenfunction of the operator, $\hat{P}_x$. The eigenvalue is $-\frac{ih}{2}$ which is the linear momentum.

Let us try for $\hat{P}_x^2$

$$\hat{P}_x = -\frac{ih}{2\pi}\frac{d}{dx} \text{ and hence } \hat{P}_x^2 = \frac{i^2h^2}{4\pi^2}\frac{d^2}{dx^2}$$

$$\hat{P}_x^2\psi = \frac{i^2h^2}{4\pi^2}\frac{d^2}{dx^2}\left(Ae^{\pi x}\right)$$

$$\hat{P}_x^2\psi = -\frac{h^2}{4}\left(Ae^{\pi x}\right) = -\frac{h^2}{4}\psi$$

So $\qquad \hat{P}_x^2\psi = P_x^2\psi$

Hence, ψ is an eigenfunction of the operator $\hat{P}_x^2$. The eigenvalue is $\frac{-h^2}{4}$.

4. Show that the linear momentum operator is Hermitian

An operator is defined to be Hermitian, if it satisfies the equation

$$\int_{-\infty}^{+\infty} \psi_n^* \hat{R} \psi_m d\tau = \int_{-\infty}^{+\infty} \psi_m \hat{R}^* \psi_n^* d\tau$$

Show that $\hat{P}_x = -\frac{ih}{2\pi}\frac{\partial}{\partial x}$ is Hermitian.

$$\int_{-\infty}^{+\infty} \psi_n^* (-\frac{ih}{2\pi}\frac{\partial}{\partial x})\psi_m dx = \int_{-\infty}^{+\infty} \psi_m (-\frac{ih}{2\pi}\frac{\partial}{\partial x})^* \psi_n^* dx$$

To show this, integrate the LHS of the equation by parts that results in RHS.

$$\int_{-\infty}^{+\infty} \psi_n^* (-\frac{ih}{2\pi}\frac{\partial}{\partial x})\psi_m dx = (-\frac{ih}{2\pi})(\psi_n^*\psi_m)\Big|_{-\infty}^{+\infty} + (\frac{ih}{2\pi})\int_{-\infty}^{+\infty} \psi_m \frac{\partial}{\partial x}\psi_n^* dx$$

$$= \int_{-\infty}^{+\infty} \psi_m (\frac{ih}{2\pi}\frac{\partial}{\partial x})\psi_n^* dx = \int_{-\infty}^{+\infty} \psi_m (-\frac{ih}{2\pi}\frac{\partial}{\partial x})^* \psi_n^* dx$$

This proof of course follows for functions which vanish at the limits.

5. Define the Unitary Operator

A unitary operator is also a linear operator and defined as

$$\int \psi_1^* \hat{U}^{-1} \psi_2 d\tau = \int \psi_2 \hat{U}^* \psi_1^* d\tau \qquad (4.5.1)$$

where the operator $\hat{U}^{-1}$ is the inverse of $\hat{U}$, such that $\hat{U}^{-1}\hat{U} = \hat{U}\hat{U}^{-1}$ where ψ_1 and ψ_2 are any two eigenfunctions of $\hat{U}$. The asterisk stands for the complex conjugate quantity.

Consider the equation

$$\hat{U}\psi = \lambda\psi \qquad (4.5.2)$$

where λ is the eigenvalue. Then

$$\hat{U}^{-1}\hat{U}\psi = \psi \qquad (4.5.3)$$

or $\qquad \hat{U}^{-1}\hat{U}\psi = \psi = \lambda\hat{U}^{-1}\psi \qquad (4.5.4)$

Inverse operator $\hat{U}^{-1}$ has the same eigenfunction as $\hat{U}$ but with reciprocal eigenvalue.

Multiplying Eq.4.5.4 with ψ^* and integrating over all space one gets

$$\int \psi^* \hat{U}^{-1} \psi d\tau = \lambda^{-1} \int \psi^* \psi d\tau \qquad (4.5.5)$$

Taking the complex conjugate of Eq.4.5.2

$$\hat{U}^* \psi^* = \lambda^* \psi^*$$

Then $\qquad \int \psi \hat{U}^* \psi^* d\tau = \lambda^* \int \psi^* \psi d\tau \qquad (4.5.6)$

Following the definition of unitary operator, the RHS of Eqs. 4.5.5 and 4.5.6 are equal so that

$$\lambda^{-1} = \lambda^* \text{ or } \lambda\lambda^* = 1$$

Thus eigenvalues of an unitary operator have modulus one.

6. Find out whether the following functions are well behaved or not.

1. $\psi = x, \quad x \geq 0$

    ψ is not a well behaved function.

    ψ does not remain constant as $x \to \infty$.

2. $\psi = x^2$

    ψ is not a well behaved function.

    ψ does not remain constant as $|x| \to \infty$.

3. $\psi = \cos x$

   ψ is a well behaved function

4. $\psi = e^{-|x|}$

   ψ is not a well behaved function

   The first derivative is not continuous at x = 0.

5. $\psi = e^{-x^2}$

   ψ is a well behaved function

6. $\psi = \sin|x|$

   ψ is not a well-behaved function

   The first derivative is not continuous at x = 0.

CHAPTER 4

Applications of Schrödinger Equation-1
(Simple systems with constant potential energy)

4.1 Particle in a One-dimensional Box

As an application of the postulates of quantum mechanics, we now discuss a simple problem that of a particle constrained to move in a one dimensional box. This problem is an excellent one because it illustrates a number of quantum mechanical principles, and at the same time shows how discrete energy levels inevitably arise, whenever a small particle is confined to a region in space.

For the sake of simplicity, a one-dimensional box will be considered. In three-dimensional box, the wave function is represented by ψ_{xyz} and in one-dimensional box by ψ_x. Since the particle is to be some sort of a realistic particle, such as an electron, our wave function must be a function that does things a real particle will do. Such a function is known as a well behaved function. This requires that it is everywhere continuous, smooth, finite, and single valued.

Let a particle be placed in a one-dimensional box shown below.

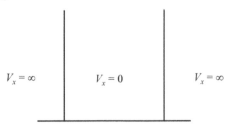

Particle in a one-dimensional box

To solve a problem in wave mechanics, it is necessary to solve the wave equation for the particular problem at hand.

$$\nabla^2 \psi_{xyz} + \frac{8\pi^2 m}{h^2}(E-V)\psi_{xyz} = 0$$

For the case of one-dimensional system, the wave equation reduces to

$$\frac{d^2\psi_x}{dx^2} + \frac{8\pi^2 m}{h^2}(E-V)\psi_x = 0$$

Chapter 4 | Applications of Schrödinger Equation-1

For the case of one-dimensional system, the wave equation reduces to

$$\frac{d^2\psi_x}{dx^2} + \frac{8\pi^2 m}{h^2}(E-V)\psi_x = 0$$

It was assumed that while the particle remained in the box, it had zero potential energy. Thus, as long as the particle remains in the box, its potential energy will be zero and the wave equation will reduce further to,

$$\frac{d^2\psi_x}{dx^2} + \frac{8\pi^2 m}{h^2}E\psi_x = 0$$

This can be simplified to,

$$\frac{d^2\psi_x}{dx^2} + \alpha^2 \psi_x = 0$$

by letting

$$\alpha^2 = \frac{8\pi^2 m}{h^2}E$$

$$\therefore \quad \frac{d^2\psi_x}{dx^2} = -\alpha^2 \psi_x$$

This is a second order differential equation whose solutions are, functions, that when differentiated twice, will give the same functions back, multiplied by a constant. The solution to the above equation was shown to be,

$$\psi_x = A\sin \alpha x + B\cos \alpha x$$

The above function is a solution to the wave equation for the particle in a one-dimensional box. (A second order differential equation will in general contain two arbitrary constants).

As such, the general solution to the differential equation gives very little information. However, we know certain restrictions that apply to this particular system. These are known as boundary conditions. For instance, since the particle must not exist outside the box, it is necessary for the wave function, ψ_x to go to zero at the walls of the box. This means that for our one-dimensional box shown in the above figure, $\psi_x = 0$ at the point x = 0. Thus we find out at the point x = 0,

$$B\cos \alpha(0) = 0$$

In order for the equality to hold good, it is obvious that the constant, B, must equal to zero.

As a result of this boundary condition, the wave function reduces to,

$$\psi_x = A\sin \alpha x$$

At the other wall it is seen that the wave function must again go to zero, and therefore at the point x = a, it is again necessary that $\psi_x = 0$. This condition offers two possible solutions. For the point x = a, the wave function becomes,

$$A \sin \alpha x = 0$$

The left side of the equation may be forced to equal to zero by letting A equal to zero. This would maintain the identity, but it would accomplish nothing towards a useful solution. Such a solution is a trivial solution. However, there is another way in which the identity may be maintained. The sine of an angle is zero at any integral multiple of π. Thus if $\alpha = \dfrac{n\pi}{a}$, where 'n' is an integer, the identity can still be satisfied. As a result of applying these boundary conditions, the wave equation for the particle now becomes,

$$\psi_x = A \sin \frac{n\pi}{a} x$$

The only term yet to be determined is the coefficient, A. This can be determined by normalizing the wave function. Since it is known that the particle must be in the box, the probability that it is in the box is unity. Knowing that this probability is represented by the square of the wave function, we can say that,

$$\int_0^a \psi_x \psi_x^* dx = 1$$

But $\quad \psi_x = A \sin \dfrac{n\pi}{a} x$ and $\alpha = \dfrac{n\pi}{a}$.

∴ $\quad \displaystyle\int_0^a A^2 \sin^2 \alpha x \, dx = 1$

∴ $\quad A = \sqrt{\dfrac{2}{a}}$

Therefore, the normalized wave function for the particle in a one-dimensional box is,

$$\psi_x = \sqrt{\frac{2}{a}} \sin \frac{n\pi}{a} x$$

It is apparent that the wave function does not have to be determined in order to find the energy of the particle,

$$\alpha^2 = \frac{8\pi^2 m}{h^2} E$$

But $\alpha = \dfrac{n\pi}{a}$ or $\alpha^2 = \dfrac{n^2\pi^2}{a^2}$

$\therefore \quad \dfrac{8\pi^2 m}{h^2} E = \dfrac{n^2\pi^2}{a^2}.$

This can be solved for energy, giving,

$$E_n = \dfrac{n^2 h^2}{8ma^2}$$

These results are represented diagrammatically Fig.4.1.1

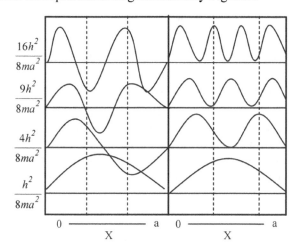

Fig. 4.1.1 Schematic drawing of E_n, ψ_n and ψ_n^2 for the case of a particle moving in a one-dimensional box. Note that ψ_n changes sign at each node while ψ_n^2 always remains positive.

4.1.1 Salient Instructive Features of the Problem

1. For the same value of the quantum number 'n', the energy is inversely proportional to the mass of the particle and square of the length of the box. Thus, as the particle becomes heavier and the box larger, the energy levels become more closely spaced. It is only when the quantity, ma^2 is of the same order as h^2, that quantized energy levels become important in experimental measurements. When dealing with dimensions of 1 g and 1 cm, the energy levels become so closely spaced that they seem to us, to be continuous. The quantum mechanical formula, therefore, gives the classical result for systems with dimensions such that, $ma^2 \gg h^2$. This is an illustration of the "correspondence principle" that states that the quantum mechanical result must become identical with the classical one in the limit

where the quantum numbers describing the system become very large, where, n^2h^2 is comparable to 'ma²', then even the macro particles need quantum mechanical treatment.

2. The second feature of the solutions to the particle in a box problem that should be pointed out is the relationship between the energy of a state and the number of nodes in the wave function. A node is a point where the wave function becomes zero. Neglecting the nodes at the end of the box, in the state n = 2 there is one node, in n = 3, two nodes and in state n, n-1 nodes. It is a general property of the wave functions that the greater the number of nodes in a wave function, the higher the energy of the corresponding state. This is shown in the above figure and is reasonable when considered along with the de Broglie relationship. The greater the number of nodes along the length of the box, the shorter the wavelength must be. According to the de Broglie relation $\lambda = \dfrac{h}{p}$, if the wavelength λ becomes shorter, the momentum and hence the kinetic energy of the particle, must be greater.

3. One of the conditions of Bohr's postulates in deciding the size of an orbit is that the angular momentum of an electron circulating in that orbit should be an integral multiple of $\hbar$. This can be explained on the basis of associating wave and particle properties to the electron. As per the postulate the angular momentum of the electron, mvr = $n\hbar$. Because the electron exhibits dual properties, the linear momentum, 'mv' can be equated to $\dfrac{h}{\lambda}$.

$$\therefore mvr = n\dfrac{h}{2\pi} \text{ where, n is an integer}$$

$$\dfrac{h}{\lambda}r = n\dfrac{h}{2\pi} \text{ and hence } 2\pi r = n\lambda$$

This clearly shows that unless the circumference of the orbit is an integral multiple of the wavelength of the wave associated with the electron, the wave destroys itself, and the electron cannot have existence. When this condition is satisfied, it results in a stationary wave confining the electron to a fixed orbit, until it gains or loses energy to go to a higher or lower orbit.

4. Another feature is the explanation for the stability of the atom and the energy of confinement, which can be done more elegantly based on the above observations.

The stability of an atom can be explained based on energy of confinement. As it is seen above, as the wavelength decreases with decrease in the size of the orbit, the energy of the electron increases. Thus, it opposes the attractive force due to the opposite charges of the nucleus and the electron. Therefore, an optimum condition is reached, at which, the attractive forces just balances the opposing forces, due to confinement energy and that state of the electron corresponds to its presence in an orbit.

This gives a better explanation for the observation why an accelerating charge like an electron while moving round the nucleus does not radiate energy and fall into the nucleus, rather than simply saying that as long as the electron is in an orbit it neither gains or loses energy.

5. An important feature of the solutions to the particle is a box problem is illustrated by integral,

$$\langle \psi_1 | \psi_2 \rangle = \frac{2}{a} \int_0^a \sin\frac{\pi x}{a} \sin\frac{2\pi x}{a} dx$$

To evaluate the integral we make the substitution,

$$y = \frac{\pi x}{a}$$

then $\langle \psi_1 | \psi_2 \rangle = \frac{2}{\pi} \int_0^\pi \sin y \sin 2y \, dy$

$$= 0$$

Whenever an integral of the type $\langle \psi_i | \psi_j \rangle$ equals to zero, then ψ_i and ψ_j are said to be orthogonal. Thus, the wave functions ψ_1 and ψ_2 are orthogonal. In fact, it can be shown that the integrals of the type $\langle \psi_i | \psi_j \rangle$, where i ≠ j, are equal to zero for the particle in a box.

The general theorem is that "the eigenfunctions of a Hermitian operator belonging to different eigenvalues of a Hermitian operator are orthogonal".

So we write, $\langle \psi_i | \psi_j \rangle = \delta_{ij}$ $\delta_{ij} = 1$ for i = j

$$\delta_{ij} = 0 \text{ for } i \neq 0$$

Where δ_{ij} is called the Kronecker delta.

6. We next inquire about some other properties of the particle in a box. Suppose we are interested in measuring the component of momentum in the 'x' direction for a set of identical systems in which the particle is known to be in the lowest energy state.

The operator for the momentum is, $\hat{P}_x = \frac{-ih}{2\pi} \frac{d}{dx}$

∴ $\hat{P}_x \psi_1 = \frac{-ih}{2\pi} \frac{d}{dx}\left(A \sin\frac{\pi x}{a}\right)$

$= \frac{-ih}{2a}\left(A \cos\frac{\pi x}{a}\right)$

It is clear that ψ_1 is not an eigenfunction of $\hat{P}_x$. Therefore, according to postulate IV, a series of measurements of $\hat{P}_x$ will not yield the same result. We must use the average value theorem to calculate the expectation value of $\hat{P}_x$. This gives,

$$\langle \hat{P}_x \rangle = \frac{\int_0^a \psi_1 \hat{P}_x \psi_1 . dx}{\int_0^a \psi_1 . \psi_1 dx} \qquad (4.1)$$

$$\langle \hat{P}_x \rangle = \frac{\int_0^a \left[\sqrt{\frac{2}{a}} \sin\frac{\pi a}{a} \left(\frac{-ih}{2\pi} \right) \frac{d}{dx} \sqrt{\frac{2}{a}} \sin\frac{\pi x}{a} \right] dx}{\int_0^a \left(\sqrt{\frac{2}{a}} \sin\frac{\pi x}{a} \right)^2 dx}$$

$$\langle \hat{P}_x \rangle = \frac{\int_0^a \left[\sqrt{\frac{2}{a}} \sin\frac{\pi a}{a} \left(\frac{-ih}{2\pi} \right) \sqrt{\frac{2}{a}} \frac{\pi}{a} \cos\frac{\pi x}{a} \right] dx}{1}$$

$$\langle \hat{P}_x \rangle = 0$$

Accordingly, the average of a large number of measurements of $\langle \hat{P}_x \rangle$ on the set of identical systems is zero.

Suppose one now considers the square of the momentum in the 'x' direction. The appropriate operator is,

$$\hat{P}_x^2 = \left(\frac{-ih}{2\pi} \frac{d}{dx} \right)^2.$$

Applying the operator we obtain,

$$\hat{P}_x \psi_1 = \left(\frac{-ih}{2\pi} \frac{d}{dx} \right)^2 \left(A\sin\frac{\pi x}{a} \right)$$

$$\hat{P}_x^2 \psi_1 = \frac{h^2}{4a^2} \left(A\sin\frac{\pi x}{a} \right) = \frac{h^2}{4a^2} \psi_1 \qquad (4.2)$$

A constant times the wave function is generated and hence the constant is the eigenvalue for the operator $\hat{P}_x^2$. Hence the square of the momentum is,

$$P_x^2 = \frac{h^2}{4a^2} = 2mE$$

$$\therefore \quad P_x = \pm\sqrt{2mE} .$$

The results calculated in equations (4.1) and (4.2) present interesting dilemma. The result of equation (4.1) indicates that the average value of P_x is zero. The result of equation (4.2) indicates that the value of P_x must be either $\pm\sqrt{2mE}$.

The apparent contradiction is resolved by considering the meaning of postulates III and IV. Since a measurement of P_x^2 always gives the result $2mE$, the momentum P_x must always be either $+\sqrt{2mE}$ or $-\sqrt{2mE}$. A single measurement of P_x will give one of these values. What the mean value postulate states is that, if one makes a large number of measurements of P_x, we end up with $P_x = +\sqrt{2mE}$ as often as $P_x = -\sqrt{2mE}$ and the average value of P_x is zero. The important point is that we never know in advance whether an experimental result will give plus or minus $\sqrt{2mE}$. It can therefore be said that an uncertainty exists in our knowledge of the momentum, and the magnitude of this uncertainty is equal to $2\sqrt{2mE}$.

In a similar manner, we can argue that if we know that the particle is in state ψ_n, the only thing that can be said about the position of the particle is that it is somewhere in the box. That is, our uncertainty in the 'x' coordinate of the particle in the length of the box, a. It is of interest to calculate the product of our uncertainties in the position and momentum of a particle in a box. This is,

$$\Delta x . \Delta p \geq a \times 2\sqrt{2mE}$$

$$\Delta x . \Delta p \geq a \times 2\sqrt{\frac{h^2}{4a^2}}$$

$\Delta x . \Delta p \geq h$ or in general n × h.

This will have its smallest value when n = 1, and thus we obtain the result that $\Delta x . \Delta p \approx h$. This is a form of the Heisenberg's uncertainty principle, which states that the simultaneous measurement of both the position and momentum of a particle cannot be made to an accuracy greater than Planck's constant '*h*'.

4.1.2 Zero Point Energy

The above conclusion is a result of the zero point energy.

Consider $\psi_x = A\sin\dfrac{n\pi x}{a}$.

In this equation although the value zero for 'n' is permitted, it is not acceptable because ψ becomes zero, as it contradicts the assumption that an electron is assumed to be always present inside the box. Therefore, the lowest kinetic energy, called the zero point energy, E_0 of an electron in a box is given by [Substitute n = 1],

$$E_0 = \frac{h^2}{8ma^2}$$

This shows that the electron inside the box is not at rest even at 0 K. Therefore, the position of the electron cannot be precisely known. Since only the mean value of the electron is known to be zero and the exact energy is not precisely known, the occurrence of zero point energy is therefore in accordance with the Heisenberg's uncertainty principle.

4.1.3 Free Particle

If the walls of the box are removed and the electron is free to move without any restriction in a field whose potential energy may be assumed to be zero then the Schrödinger's wave equation and its solutions are given by,

$$\frac{d^2\psi_x}{dx^2} + \frac{8\pi^2 m}{h^2} E\psi_x = 0$$

and $\psi_x = A\sin\alpha x + B\cos\alpha x$, where $\alpha^2 = \frac{8\pi^2 m}{h^2} E$.

The arbitrary constants A, B and α^2 can now have any value. Then the energy,

$$E_0 = \frac{h^2}{8ma^2} \text{ is not quantized.}$$

Thus, when an electron is bound in a system, it has quantized energy levels, and it leads to discrete spectrum. On the other hand, a free particle (electron) moving without any restriction has the continuous energy spectrum. This qualitatively explains the occurrence of continuum in the atomic or molecular spectra. But, on ionization, an electron lost by an atom or molecule is free to move without any restriction.

4.2 The Particle in a Three Dimensional Box

For the particle in a three dimensional box, the wave function will be a function of all three space coordinates. The wave equation for such a particle moving in a region of zero potential energy is,

$$\nabla^2 \psi_{xyz} + \frac{8\pi^2 m}{h^2} E\psi_{xyz} = 0$$

Chapter 4 | Applications of Schrödinger Equation-1

This is a partial differential equation containing three variables, and the standard approach to the solution of such an equation, is about the same, as that used to separate the time and space parts of the time dependent wave equation.

First, it is assumed that the total wave function can be represented as a product of wave functions. For a particle in three-dimensional box, it is then assumed that,

$$\psi_{xyz} = x_x \cdot y_y \cdot z_z$$

where x_x represents a wave function that depends on the variable 'x' only, y_y represents a wave function that depends on the variable on 'y' only and so on.

Therefore, we obtain

$$\left(\frac{\partial^2}{\partial x^2} + \frac{\partial^2}{\partial y^2} + \frac{\partial^2}{\partial z^2}\right) x_x y_y z_z + \frac{8\pi^2 m}{h^2} E x_x y_y z_z = 0$$

Since the operator $\frac{\partial^2}{\partial x^2}$ has no effect on 'y' and 'z' and the operator $\frac{\partial^2}{\partial y^2}$ has no effect on 'x' and 'z' etc., the wave equation after division by $x_x y_y z_z$ may be rearranged to give,

$$\frac{1}{x_x}\frac{\partial^2 x_x}{\partial x^2} + \frac{1}{y_y}\frac{\partial^2 y_y}{\partial y^2} + \frac{1}{z_z}\frac{\partial^2 z_z}{\partial z^2} = -\frac{8\pi^2 m}{h^2}E$$

It is to be noted that each term on the left side is a function of one variable and the sum of these terms is the constant, $-\frac{8\pi^2 m}{h^2}E$. If we keep the variables 'y' and 'z' constant and allow 'x' to vary, it is seen that the sum of the three terms is still the same constant. Such a situation, can only exist, if the term,

$$\frac{1}{x_x}\frac{\partial^2 x_x}{\partial x^2}$$

is independent of 'x', and therefore itself is a constant. The same argument will apply equally well to the 'y' and 'z' terms. Thus, each variable is seen to be independent of the other variables, and we have succeeded in separating the variables.

Now, if the constants are represented by $-\alpha_x^2$ for 'x' term, $-\alpha_y^2$ for the 'y' term and $-\alpha_z^2$ for the 'z' term, the following three differential equations are obtained.

$$\frac{1}{x_x}\frac{d^2 x_x}{dx^2} = -\alpha_x^2$$

$$\frac{1}{y_y}\frac{d^2 y_y}{dy^2} = -\alpha_y^2$$

$$\frac{1}{z_z}\frac{d^2 z_z}{dz^2} = -\alpha_z^2$$

$$\therefore \quad \alpha_x^2 + \alpha_y^2 + \alpha_z^2 = \frac{8\pi^2 m}{h^2}E$$

Thus, each degree of freedom can make its own contribution, such that,

$$\alpha_x^2 = \frac{8\pi^2 m}{h^2}E_x$$

$$\alpha_y^2 = \frac{8\pi^2 m}{h^2}E_y$$

$$\alpha_z^2 = \frac{8\pi^2 m}{h^2}E_z$$

Now that the variables have been separated, it is necessary to solve each of the equations. In this particular problem, all three of the resulting equations are of the same form. Thus, the solution of one is sufficient to demonstrate the method. If the equation in 'x' is used as an example, it is seen that on rearrangement, it is of the exact form as the wave equation we have just solved for the one-dimensional box.

$$\frac{d^2 x_x}{dx^2} + \alpha_x^2 x_x = 0$$

Then the normalized solution for the above equation is,

$$x_x = \sqrt{\frac{2}{a}}\sin\frac{n_x \pi}{a}x$$

and an analogous solution would be obtained for the 'y' and 'z' equations.

Since, $\quad \psi_{xyz} = \sqrt{\frac{8}{abc}}\sin\frac{n_x \pi}{a}x.\sin\frac{n_y \pi}{b}y.\sin\frac{n_z \pi}{c}z$

it is significant to note that there is a quantum number for each degree of freedom. This same idea was emphasized in the Sommerfeld's quantization of hydrogen atom, but here the quantization is a natural consequence of the mathematics.

As in the previous case $\alpha = \dfrac{n\pi}{x}$, and here

$$\alpha_x = \frac{n_x \pi}{a}$$

$$\alpha_y = \frac{n_y \pi}{b} \quad \text{and}$$

$$\alpha_z = \frac{n_z \pi}{c}$$

Hence, $$E_x = \frac{h^2}{8m}\left(\frac{n_x^2}{a^2}\right)$$

$$E_y = \frac{h^2}{8m}\left(\frac{n_y^2}{b^2}\right)$$

$$E_z = \frac{h^2}{8m}\left(\frac{n_z^2}{c^2}\right)$$

But $$E = E_x + E_y + E_z = \frac{h^2}{8m}\left(\frac{n_x^2}{a^2} + \frac{n_y^2}{b^2} + \frac{n_z^2}{c^2}\right)$$

Here, again, it is seen that the energy of the particle is quantized.

(This might lead one to wonder at the success of the classical approach to the mechanics of atoms and molecules as found in the kinetic theory of gases. Actually, no conflict exists between the two approaches. If quantum numbers and containers of reasonable size are chosen, it is found that the separation of energy levels is so small that the energy distribution will essentially be continuous.)

Degeneracy

For a complete description of the energy states of a particle in a three dimensional box, we see that it is necessary to consider three quantum numbers. This, of course, is what one should expect. The idea of quantum numbers in atomic spectra, for instance, came from an attempt to understand the positions of the spectral lines and the energies they represent. The observation of new lines necessarily led to a new quantum number, which could be associated with the corresponding new energy levels. Thus, we are prone to conclude that each quantum number represents a contribution to the energy of the system. However, it is frequently found that for various reasons, a particular set of quantum numbers may not be unique in defining the energy of a particle.

To consider again, if a = b = c,

Quantum Chemistry

$$E = E_x + E_y + E_z = \frac{(n_x^2 + n_y^2 + n_z^2)}{8ma^2} h^2$$

where, n_x or n_y or n_z = 1, 2, 3, 4,....... The occurrence of three quantum numbers n_x, n_y and n_z is characteristic of a three-dimensional problem in wave mechanics. The zero point energy is three times that observed in the one-dimensional box, there being a part associated with each of the three coordinates.

Since each state is characterized by three quantum numbers, it is possible to construct several excited states of the same energy. For example, there are three independent states having the quantum numbers (2, 1, 1), (1, 2, 1) and (1, 1, 2) for the same energy. All these three states have the same energy $\frac{6h^2}{8ma^2}$. Such a state is threefold degenerate or triply degenerate. In the Fig. 4.2.1 below, is shown a few energy levels illustrating the degeneracy and the zero point energy.

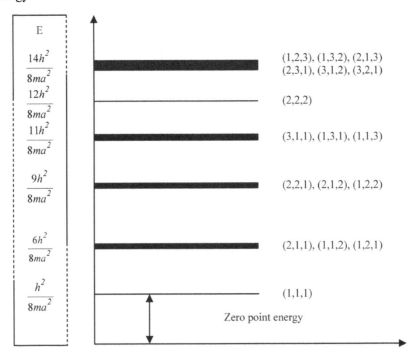

Fig. 4.2.1 Degeneracy of various states.

Chapter 4 | Applications of Schrödinger Equation-1

Note.

Although the value of zero for 'n' is permitted for a wave function, $\psi_n = D\sin\dfrac{n\pi x}{a}$ in the case of a particle in a one dimensional box, it is not acceptable because then ψ_n becomes zero. However, the electron is assumed always present in the box. Therefore, the lowest K.E. called the zero point energy, of an electron in a one dimensional box is given by,

$$E_{ZPE} = \frac{h^2}{8ma^2}$$

This shows that the electron inside the box is not at rest at 0 K. Therefore, the position of the electron is not precisely known.

In the case of a three dimensional box the zero point energy is three times to that of $\dfrac{h^2}{8ma^2}$ as there being, a part associated with each of the three coordinates.

The wave functions of the three members of the above triply degenerate levels are different.

Consider a slight distortion of the cube by 'da' along the 'x' axis. For the state (2, 1, 1) the energy of the electron in the undistorted cube may be given as,

$$E = E_x + E_y + E_z$$

$$= \frac{4h^2}{8ma^2} + \frac{h^2}{8ma^2} + \frac{h^2}{8ma^2} = \frac{6h^2}{8ma^2}$$

The new energy on distortion along the 'x' axis is,

$$E + d_E = E_x + dE_x + E_y + E_z$$

$$= \frac{4h^2}{8ma^2} - \frac{h^2}{ma^3}da + \frac{h^2}{8ma^2} + \frac{h^2}{8ma^2} = \frac{6h^2}{8ma^2} - \frac{h^2}{ma^3}da$$

whereas the new energy for the other two states, i.e.(1,2,1) and (1,1,2) is ,

$$E + d_E = \frac{6h^2}{8ma^2} - \frac{h^2}{4ma^3}da$$

Thus, the initial threefold degenerate levels are split on distortion of the cube into a non-degenerate level and doubly degenerate levels.

It is a common phenomenon in chemistry that the electronic degeneracy is removed on a slight distortion of a system. This is analogous to the "Jahn-

Teller effect" which states that, "in a nonlinear molecule, where electronic degeneracy occurs, there always exists a vibrational mode, which can remove the degeneracy". The molecule can, therefore, spontaneously distort from its most symmetric configuration, until it assumes a configuration of lower symmetry and lower energy.

4.3 The Structure of Matter

The structure of matter can be analyzed according to increasingly more fundamental levels of organization. Table 4.3.1 shows this in an elementary way. A piece of metal, for example, is made of atoms, which are kept together by a chemical force that does not contain very high energies. Using electron volts as a measure of energy, we would need only a few eV to separate an atom from the piece of metal. If we look at one of the atoms, we see that it consists of a nucleus surrounded by the electric force, which is somewhat stronger than the chemical forces. From ten to few hundred electron volts are needed to tear a few electrons from an atom. The size of an atom is about an angstrom.

Table 4.3.1: Fundamental Structural Levels of Matter

	Constituent particles	Bonding force	Relevant energies, eV
Matter	Atoms	Chemical forces	1
Atom	Electrons, Nucleus	Electric force	10-1000
Nucleus	Protons, Neutrons	Nuclear force	10^6
Nucleon	Quarks	Strong force	10^9

Now let us look at the nucleus, which is considerably smaller, of the order of a pico-meter. The nucleus consists of protons and neutrons kept together by the nuclear force; a force that is much larger than the chemical and electric forces. Energies in the range of about 10 MeV are required to tear a proton or a neutron from a nucleus.

We know today that protons and neutrons, the so-called nucleons, are themselves not elementary but probably consist of three elementary particles that carry the unhappy name 'quark'. One wishes, a better sounding name had been chosen, such as 'parton' for example, but 'quark' has been taken. The forces that keep the quarks within the nucleon are again much stronger. The size of the nucleon is about 10^{-13} cm, and the effects of the forces involve energies of the order of one billion eV(GeV). The nuclear force that binds protons and neutrons in the nucleus is understood today because of the 'strong force', which keeps quarks together.

This analysis brings us to the two related concepts that play an important role in many theories of the history of the universe: the 'quantum ladder' and the 'conditional elementarity' of particles. We may distinguish three different realms in nature, three levels on the quantum ladder as shown in Table 4.3.2.

Table 4.3.2: The quantum ladder

Subject	Energy range, eV	Main location
Atomic and molecular realm Chemistry, optics, materials, biology, complexity, organization, order-disorder	up to 1000	Earth and planets of other stars
Nuclear realm Radio activity, Nuclear reactions, fission, fusion	$10^5 - 10^7$	Interior of stars
Sub nuclear realm Antimatter, mesons, heavy electrons, short lived entities, quarks	10^8 - ?	Big-bang, neutron stars, unknown

The first is the atomic realm, which includes the world of atoms, their interactions, and the structures that are formed by them, such as molecules, liquids and solids, gases and plasmas. This realm includes all phenomena of atomic physics, chemistry, and in a certain sense, biology. The energy changes taking place in this realm are a few eV. If these exchanges are below 1 eV, such as in the collisions between air molecules in the room, then even atoms and molecules can be regarded as elementary particles, That is they retain the 'conditional elementarity' depicted in Table 4.3.3, as they keep their identity during any collisions or in other processes at these low energy exchanges. If one goes to higher energy exchanges, say 10,000 eV, atoms and molecules will decompose into nuclei and electrons. At this level, the nuclei and electrons must be considered as elementary. The structures and processes of the first rung of the quantum ladder are found on earth, on planets and on the surface of stars.

Table 4.3.3: Conditional Elementarity of Particles

Particles	Energy limit
Molecules, Photons	< 0.1 eV
Atoms, Photons	< 1.0 eV
Nuclei, Electrons	< 10^4 eV
Protons, Neutrons, Electrons, Nutrinos, Photons	<10^9 eV
Quarks, Electrons, Muons, Tauons, W, Z, Gluons, Photons	> 10^9 eV

64 Quantum Chemistry

The next rung is the nuclear realm. Here the energy changes are much higher, of the order of mega(million) electron volts(MeV). As long as we are dealing with phenomena in the atomic realm, such amounts of energy are unavailable, and the nuclei remain inert and do not change. However, if one applies energies of millions of eV, nuclear reactions like fission and fusion, and the processes of radio activity occur, Our elementary particles then will be protons and neutrons, together with electrons.

In addition, radioactive processes produce neutrinos, particles that have no detectable mass or charge. In the universe, energies at this level are available in the centers of stars and in star explosions. Indeed, the energy radiated by the stars is produced by the nuclear reactions. The natural radioactivity we find on earth is the long-lived remnant of the time when earthly matter was expelled into space by a supernova explosion.

The third rung of quantum ladder is the sub-nuclear realm. Here we are dealing with energy exchanges of many giga electron volts(GeV), or billions of electron volts. We encounter excited nucleons, new types of particles such as mesons, heavy electrons, quarks, gluons, and antimatter in large quantities. Gluons are the quanta of the strong force that keeps quarks together. As long as we are dealing with the atomic or nuclear realm, these new types of particles do not occur and the nucleon remains inert. But at sub-nuclear energy levels, nucleons and mesons appear to be composed of quarks, so that quarks and gluons figure as elementary particles.

It is an interesting question whether the elementary particles established so far are indeed truly elementary. It may well be that they are also conditional and that the list has to be extended further and further.

4.4 Factors Influencing Color

The wave mechanical treatment of an electron in a box gives rise to large number of discrete energy levels. On suitable excitation, the electron may undergo transition from one level to another. The transition energy for the transition $\psi_n \rightarrow \psi'_n$ is given by,

$$\Delta E = \frac{n'^2 - n^2}{8ma^2} h^2$$

Therefore, the frequency of transition obtained through Bohr's relation,

$$\nu = \frac{\Delta E}{h} \text{ is given by}$$

$$\nu = \frac{n'^2 - n^2}{8ma^2} h$$

Chapter 4 | Applications of Schrödinger Equation-1

or the wavelength of the transition is,

$$\lambda = \frac{c}{\nu} = \frac{8mca^2}{\left(n'^2 - n^2\right)h}$$

This relation shows that the longer the length of the box 'a', the longer the wavelength at which the optical transition in such a system occurs. Thus, by suitably adjusting the length of the box, the wavelength of an electronic transition can be made to appear in the visible range of the spectrum (roughly from 4000 - 8000 $\overset{0}{A}$). This system then becomes colored. However, it should be emphasized that transition between any two levels is not always permissible, because the transition probability for an electric dipole transition (which is the most important cause of light absorption) depends on the magnitude of transition dipole moment integrals defined as,

$$e\int \psi_{n'} \hat{x} \psi_n dx$$

Here it will be shown, under what condition the transition dipole moment integral is different from zero for an electron in a box.

Consider the integral

$$\int_0^a \psi_{n'} \hat{x} \psi_n dx = \frac{2}{a}\int_0^a \sin\frac{n'\pi}{a}x.\hat{x}.\sin\frac{n\pi}{a}x.dx$$

Letting $\frac{\pi x}{a} = \theta$, $x = \frac{a\theta}{\pi}$ and $dx = \frac{a}{\pi}d\theta$

we may write

$$\int_0^a \psi_{n'} \hat{x} \psi_n dx = \frac{2a}{\pi^2}\int_0^\pi \sin n\theta.\sin n'\theta.\theta d\theta$$

$$= \frac{2a}{\pi^2}\left[\frac{1}{2}\int_0^\pi \cos(n-n')\theta\theta d\theta - \frac{1}{2}\cos(n+n')\theta\theta d\theta\right]$$

$$= \frac{a}{\pi^2}[I_1 - I_2]$$

Where $I_1 = \int_0^\pi \cos(n-n')\theta.\theta d\theta$

$$= \left[\theta.\frac{\sin(n-n')\theta}{(n-n')}\right]_0^\pi - \int_0^\pi \frac{\sin(n-n')\theta}{(n-n')}.d\theta$$

$$= \left[0 + \frac{-2}{(n-n')^2}\right] = \frac{-2}{(n-n')^2}, \quad \text{if } (n-n') \text{ is an odd number}$$

$$= 0 \quad \text{if } (n-n') \text{ is an even number}$$

and

$$I_1 = \int_0^\pi \cos(n+n')\theta.\theta.d\theta$$

$$= \left[0 + \frac{-2}{(n+n')^2}\right] = \frac{-2}{(n+n')^2}, \quad \text{if } (n+n') \text{ is an odd number}$$

$$= 0 \quad \text{if } (n+n') \text{ is an even number}$$

This simple argument shows that the transition dipole integral $\int \psi_n . \hat{x} \psi_n dx$ does not vanish if $n \pm n'$ is an odd number. Then the selection rule for the electron dipole transition in this system may be stated as follows:

A transition between a pair of states is possible if the sum or difference in quantum numbers is an odd number. If the sum or difference is an even number, the transition is said to be strictly forbidden.

Note:

Let ψ_1 and ψ_2 be two orbitals of an atom, with corresponding energies E_1 and E_2. If an incoming electromagnetic wave is to excite an electron from ψ_1 to ψ_2, the electric field must bring about a displacement of charge. The displacement must give rise to a dipole moment, however transitory; otherwise, it will have no effect. We can put this idea in another way owing to its occurrence during the transition of an electron between two different orbitals, a transition dipole moment must occur. The recipe for calculating the value of the transition dipole moment, 'd' is

$$d = \int \psi_2 \hat{\mu} \psi_1 dv$$

$\hat{\mu}$ is the dipole moment operator. Fortunately, this operator does not directly involve partial differentials. In fact it is precisely the same as we defined earlier, i.e., $\hat{\mu} = e \times r$. We can make some useful deductions about d without a great deal of effort.

The atomic orbitals are of odd or even parity. Especially 's' and 'd' orbitals are even, and 'p' orbitals are odd. Now suppose ψ_1 is a 1s, ψ_2 is a 2s orbital, and the electric field of the wave oscillate in the 'x' direction only. This field can only interact with the 'x' component of the dipole moment, thus we have,

$$dx = \int \psi_{2s} \hat{\mu} \psi_{1s} dv$$

$$= -e \int \psi_{2s} \hat{\mu} \psi_{1s} dv$$

In terms parity the integral has the form,

$$\psi_{2s}.x.\psi_{1s} = even \times odd \times even \equiv odd .$$

If we integrate an odd function over all space, for every region where the function is positive, there will be an analogous region where it is negative. Therefore, the overall contribution of the positive and negative regions will cancel and the integral will be zero. This can be summarized as follows:

$$\int (odd \quad function) dv = 0 \quad \text{and}$$

$$\int (even \quad function) dv \neq 0 .$$

Because the dipole moment is of odd parity we must have either ψ_1 as odd and ψ_2 as even or vice versa. This follows from the fact that

Even × odd × odd = even

Odd × odd × even = even

Therefore, the Laporte selection rule can be summarized as

$$g \leftrightarrow u; \quad g \not\leftrightarrow g; \quad u \not\leftrightarrow u; .$$

4.5 Tunneling in Quantum Mechanics

Tunneling is the quantum-mechanical effect of transitioning through a classically-forbidden energy state. It can be generalized to other types of classically-forbidden transitions as well.

Consider rolling a ball up a hill. If the ball is not given enough velocity, then it will not roll over the hill. This scenario makes sense from the standpoint of classical mechanics, but is an inapplicable restriction in quantum mechanics, simply because, quantum mechanical objects do not behave like classical objects such as balls. On a quantum scale, objects exhibit wavelike behavior. For a quantum particle moving against a potential energy "hill", the wave function describing the particle can extend to the other side of the hill. This wave represents the probability of finding the particle in a certain location, meaning that the particle has the possibility of being detected on the other side of the hill. This behavior called 'tunneling' is, as if the particle has 'dug' through the potential hill. As this is a quantum and non-classical effect, it can generally only be seen in nanoscopic phenomena, where the wave behavior of particles is more pronounced.

Availability of states is necessary for tunneling to occur. In the above example, the quantum mechanical ball will not appear *inside* the hill because there is no available "space" for it to exist, but it can tunnel to the other side of the hill, where there is free space. Analogously, a particle can tunnel through the barrier, when there are states available within the barrier. The wave function describing a particle, only expresses the probability of finding the particle at a location assuming a free state exists.

One example may clarify how the "new" and "old" mechanics differ, namely the way George Gamow in 1928 explained alpha-radioactivity. The atomic nucleus experiences opposing forces: the *strong nuclear force* holding its particles together must overcome the *electric repulsion* between positive protons sharing the nucleus, which tries to *break it up*. The nuclear force wins out at short distances, that is why nuclei exist at all. But it falls off rapidly with distance and at far away the electric repulsion dominates. Consider a proton inside the nucleus. If something moves it a short distance away, the nuclear force pulls it right back, but if it somehow got far enough, the electric repulsion would push it away, never to return. An example is nuclear fission, possible in heavy nuclei of plutonium or uranium-235. Here, the nucleus contains, so many protons trying to push it apart with their electric repulsion. When an extra neutron is allowed to be pulled into the nucleus, it releases and adds a modest amount of energy that makes the entire nucleus break up into two positively charged chunks. These are separated far enough that they never come back again. Instead, electric repulsion pushes them apart even more and releases a great amount of energy.

Such nuclei, and heavy nuclei close to them in mass, are all on the brink of instability. Even without externally added energy, they find a way to get rid of some of their disruptive positive charge. The forces on protons inside these nuclei resemble those on a bunch of marbles inside the "crater" of a volcano-shaped surface with smooth sloping sides outside, but a moderately deep crater on top. The outline of the "mountain" can be viewed as representing the total force on protons in a nucleus. Inside the crater the attraction predominates, holding the protons together, while outside it the repulsion predominates, pushing them away. In the analogy of marbles inside a crater, if a marble could somehow get to the outside, say by carving **a tunnel** through the wall of the crater, this repulsion would make it roll away and it would release energy. Newtonian mechanics provides no such tunnels and the proton is imprisoned inside the crater for eternity. According to quantum mechanics, however, the proton's location is determined by a *spread-out wave function*. That wave is highest inside the "crater" of the nucleus, and if the proton materializes there, it stays trapped. The "if" clause here just helps one to imagine the process differently in quantum mechanics. If materialization of a process like this is not observable it amounts to saying it does not exist. However, the fringes of the wave extend further out, and it always has a finite, though very small, presence beyond the crater, giving a finite chance for the proton to materialize on the outside and escape. It is as if quantum laws gave it a tiny chance to "tunnel" through the barrier to the slope outside.

Systems with discontinuity in the Potential Field

Consider an electron of mass m_e moving in one dimension in the direction of positive x-axis in potential field defined by the Fig.4.5.1.

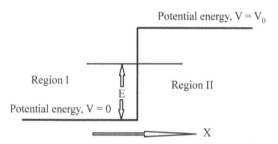

Fig. 4.5.1 A simple potential barrier.

$$V = 0 \quad \text{for} \quad x < 0$$
$$V = V_0 \quad \text{for} \quad x > 0$$

The Schrödinger equations for regions I and II are respectively

$$\frac{\partial^2 \psi_1}{\partial x^2} + \frac{8\pi^2 m_e}{h^2} E \psi_1 = 0 \tag{4.5.1}$$

$$\frac{\partial^2 \psi_2}{\partial x^2} + \frac{8\pi^2 m_e}{h^2} (E - V_0) \psi_2 = 0 \tag{4.5.2}$$

Where ψ_1 and ψ_2 are the wave functions of the particle in regions I and II respectively.

Suppose, the energy is such that $0 < E < V_0$ and let

$$k^2 = \frac{8\pi^2 m_e}{h^2} E \quad \text{and} \quad k_1^2 = \frac{8\pi^2 m_e}{h^2} (V_0 - E).$$

The appropriate wave functions for the two regions will clearly follow from the above two equations with the following solutions, respectively.

$$\psi_1 = A \exp(ikx) + B \exp(-ikx) \tag{4.5.3}$$

where A and B are arbitrary constants and

$$\psi_2 = C \exp(-k_1 x) + D \exp(+k_1 x) \tag{4.5.4}$$

Where C and D are also arbitrary constants.

The constant D has to be set to zero because, ψ_2 must vanish at infinity.

Thus $\psi_2 = C\exp(-k_1 x)$.

In Eq.4.5.3 the first term $\exp(ikx)$ is an eigenfunction of the linear momentum operator $\dfrac{-ih}{2\pi}\dfrac{\partial}{\partial x}$ with the eigenvalue, $\dfrac{kh}{2\pi}$, while the second term, $\exp(-ikx)$ has the eigenvalue $-\dfrac{kh}{2\pi}$. This suggests that the first term represents a wave travelling in the positive x-direction, i.e. the incident beam, and the second term represents a wave travelling in the negative x-direction, i.e. the reflected beam. Both the functions ψ and $\dfrac{\partial \psi}{\partial x}$ are continuous and the boundary conditions at x = 0 are therefore $\psi_1 = \psi_2$ and $\dfrac{\partial \psi_1}{\partial x} = \dfrac{\partial \psi_2}{\partial x}$

which give the following equations

$$A + B = C \tag{4.5.5}$$

$$A - B = -\dfrac{Ck_1}{ik} \tag{4.5.6}$$

or $\qquad A = \dfrac{C\left(1 - \dfrac{k_1}{ik}\right)}{2}$ and $B = \dfrac{C\left(1 + \dfrac{k_1}{ik}\right)}{2} \tag{4.5.7}$

Therefore, $\qquad \dfrac{B}{A} = \dfrac{k_1 + ik}{k_1 - ik} = \dfrac{(k_1 + ik)^2}{k_1^2 + k^2} = \alpha \tag{4.5.8}$

The intensities of the reflected and incident beams are in the ratio $|B|^2 : |A|^2$, but from Eq.4.5.8, $|\alpha|^2 = \alpha\alpha^* = 1$.

Hence, the intensity of the reflected beam equals that of the incident beam. That the wave function in Eq. 4.5.2 in the region II decays exponentially is indicative of the particles suffering almost total reflection in the region I.

However, there is a small but finite probability of particles being transmitted in the region II, which is not predictable from classical mechanics. We define the transmission coefficients by $\left|\dfrac{C}{A}\right|^2$ and it follows from Eq. 4.5.7 that

$$\left|\dfrac{C}{A}\right|^2 = \dfrac{2ik}{ik - k_1}\dfrac{2ik}{ik + k_1} = \dfrac{4k^2}{k_1^2 + k^2}$$

Substituting for k_1 and k

$$\left|\frac{C}{A}\right|^2 = \frac{4E}{V_0} \neq 0$$

This transmission coefficient is not zero unless the potential energy of the barrier is infinite. This is the basis for the 'tunneling effect' observed in quantum mechanics.

Hydrogen transfer reaction

As an example, take the hydrogen transfer reactions, especially in low temperatures. Tunneling effect is very important in these reactions. Consider a reaction

$$AH + B = A + BH.$$

Since A and B are heavier than H-atoms, we may assume that hydrogen moves between two centers which remain at a fixed distance. We represent the proton wave by the Schrödinger's equation and the proton transfer by the incidence of such a wave on the energy barrier as depicted in the following figure, which shows the variation of 'ψ' with distance superimposed on particular barrier. The wave equation is

$$\frac{\partial^2 \psi}{\partial x^2} + \frac{8\pi^2 m_H}{h^2}\left(E - V_{(x)}\right)\psi = 0$$

where m_H is the mass of the hydrogen atom, $V_{(x)}$ is the potential energy of the barrier as a function of 'x'. On substituting for $V_{(x)}$, the resulting equation can be solved in simple cases for ψ.

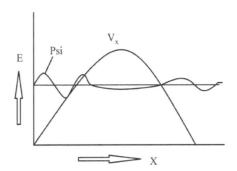

Fig. 4.5.2 Incidence of a portion wave on a barrier V(x) with energy E(E < V). Variation of ψ (Psi) with 'x' is shown.

For proton energies lower than the barrier height, the solutions show that besides the reflected wave, there is also a transmitted proton wave. It means, there is also a finite probability that proton will tunnel through the barrier.

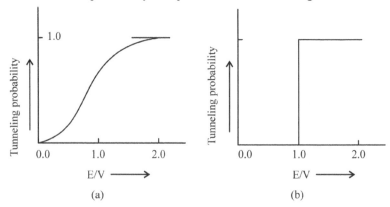

Fig. 4.5.3 (a) The tunneling probability against energy E of proton expressed as E/V for a parabolic barrier (b) The classical probability of crossing the barrier as a function of E/V.

This probability rises with the energy of the tunneling particle. Whereas the classical theory predicts that, a particle can cross a potential energy barrier only if its energy is equal to or greater than the barrier height.

It should be noted, in quantum mechanics, that all protons with energy greater than the barrier height (i.e. $(E/V)>1$) will not cross the barrier. The curve in the figure shows that the tunneling probability is less than unity even for them, indicating that partial reflection occurs.

4.6 The Rigid Rotor

Consider a system of two spherical masses m_1 and m_2 at fixed distances r_1 and r_2, respectively, from the centre of the mass of the system. The distance between the centers of the particles is assumed to have a constant value r_0, where $r_0 = r_1 + r_2$.

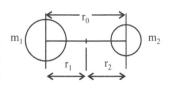

We refer to the above system as rigid rotor because the distance between the two particles is fixed and the system could rotate only about a fixed axis. In a way, the rigid rotor is an idealized case of a diatomic molecule except that in the latter the masses m_1 and m_2 (the atoms) can of course vibrate, so r_0 varies slightly but can be taken as equal to the equilibrium distance (r_e) in a real diatomic molecule. We shall consider this example in some detail since it serves as an introduction to the way in which certain important functions arise.

Chapter 4 | Applications of Schrödinger Equation-1

The theory of such, rigid rotor in space is useful in dealing with rotational spectra of diatomic molecule.

Let the distance of m_1 from the centre of gravity be r_1 and distance of m_2 be r_2, then

$$m_1 r_1 = m_2 r_2 \text{ and } r_1 + r_2 = r_0 \quad (4.6.1)$$

where,

$$r = \frac{m_r r_0}{m_1 + m_2}, \quad r_2 = \frac{m_1 r_0}{m_1 + m_2} \quad (4.6.2)$$

The kinetic energy of rotation of the atoms joined together by a link is given by

$$\text{K.E.} = \left(\frac{1}{2}\right) m_1 v_1^2 + \left(\frac{1}{2}\right) m_2 v_2^2 \quad (4.6.3)$$

where v_1 and v_2 are linear velocities of masses m_1 and m_2 respectively. Since, r_1 and r_2 are assumed to remain unchanged during rotation about the centre of gravity, one can write

$$\text{K.E.} = \left(\frac{1}{2}\right) m_1 w^2 r_1^2 + \left(\frac{1}{2}\right) m_2 w^2 r_2^2 = \left(\frac{1}{2}\right) w^2 I$$

where I denotes the moment of inertia of the system. The equation shows that the K. E. of the system is the same as that of a single particle of mass 'I' moving on the surface of a sphere of unit radius.

The Schrödinger equation takes the general form

$$H \psi = E \psi \text{ or } (T+V) \psi = E\psi \quad (4.6.4)$$

and since no forces are assumed to act on the rigid rotor

$v = 0$. Therefore $T\psi = E\psi$

In spherical polar coordinates, T takes the form

$$\frac{-h^2}{8\pi^2 m} \left[\frac{1}{r^2} \frac{\partial \left(r^2 \partial\right)}{\partial r \partial r} + \frac{1}{r^2 \sin\theta} \frac{\partial}{\partial \theta} \left(\sin\theta \frac{\partial}{\partial \theta} \right) + \frac{1}{r^2 \sin^2\theta} \frac{\partial}{\partial \varphi^2} \right] \quad (4.6.5)$$

However, for the rigid rotator we can replace 'm' by the moment of inertia 'I' and 'r' by unity and so the Schrödinger equation becomes

$$\frac{1}{\sin\theta} \frac{\partial}{\partial \theta}\left(\sin\theta \frac{\partial \psi}{\partial \theta} \right) + \frac{1}{\sin^2\theta} \frac{\partial \psi}{\partial \varphi^2} + \frac{8\pi^2 IE\psi}{h^2} = 0 \quad (4.6.6)$$

This is a differential equation with two independent variables (θ and φ) and we shall make a very common assumption at this stage, namely, that the function 'ψ' involving both θ and φ can be written as a product of two functions each involving one variable only,

$$\psi(\theta, \phi) = Y(\theta) Z(\phi) \qquad (4.6.7)$$

substitution in previous equation gives

$$\frac{Z}{\sin\theta}\frac{\partial}{\partial\theta}\left(\sin\theta\frac{\partial Y}{\partial\theta}\right) + \frac{Y}{\sin^2\theta}\frac{\partial Z}{\partial\phi^2} + \beta YZ = 0 \qquad (4.6.8)$$

where $\quad \beta = \dfrac{8\pi^2 IE}{h^2}$

Multiplying Eq.4.6.6 by $\sin^2\theta/YZ$ gives

$$\frac{\sin\theta}{Y}\frac{\partial}{\partial\theta}\left(\sin\theta\frac{\partial Y}{\partial\theta}\right) + \beta\sin^2\theta = \frac{-1}{Z}\frac{\partial^2 Z}{\partial\phi^2} \qquad (4.6.9)$$

The above equation must be valid for all values of θ and φ. This can only occur if both sides are separately equal to the same constant, say m^2, for convenience. That is

$$\frac{\sin\theta}{Y}\frac{d}{d\theta}\left(\sin\theta\frac{dY}{d\theta}\right) + \beta\sin^2\theta = m^2$$

and

$$\frac{-1}{Z}\frac{d^2 Z}{d\phi^2} = m^2 \qquad (4.6.10)$$

Solutions:

Consider the equation

$$\frac{d^2 Z}{d\phi^2} + m^2 Z = 0$$

Which has solutions

$$Z = C\exp(\pm im\phi)$$

where C is some constant to be determined. Since the rigid rotor takes up an identical configuration every time φ increases by 2π,

$$Z(\phi) = Z(\phi + 2\pi) \text{ and } Z(0) = Z(0 + 2\pi)$$

But $\quad Z(0) = C\ \exp(0)$

So $\quad Z(2\pi) = C\ \exp.(\pm 2\pi\ im)$

Chapter 4 | Applications of Schrödinger Equation-1

Since $Z(2\pi) = C$, it follows that

$$\exp(\pm 2\pi im) = 1$$

which is only true, if 'm' is zero or an integer.

The constant C is easily determined by normalization of Z i. e.

$$N^2 \int_0^{2\pi} ZZ^* d\varphi = 1 \qquad (4.6.11)$$

$$N^2 \int_0^{2\pi} (\cos m\varphi + i\sin m\varphi)(\cos m\varphi - i\sin m\varphi) d\varphi = 1$$

$$N^2 \int_0^{2\pi} d\varphi = 1$$

$$N = \left(\frac{1}{2\pi}\right)^{1/2}$$

Hence the normalized functions are

where $m = 0, \pm 1, \pm 2, \ldots\ldots$ (4.6.12)

Now consider the equation for $Y(\theta)$, which after multiplying through by $Y/\sin^2\theta$ and rearranging becomes

$$\frac{1}{\sin\theta}\frac{d}{d\theta}\left(\sin\theta\frac{dY}{d\theta}\right) + \left(\beta - \frac{m^2}{\sin^2\theta}\right)Y = 0 \qquad (4.6.13)$$

$$\frac{1}{\sin\theta}\left(\sin\theta\frac{d^2Y}{d\theta^2} + \cos\frac{dY}{d\theta}\right) + \left(\beta - \frac{m^2}{\sin^2\theta}\right)Y = 0$$

(For details refer appendix 8.11)

$$\frac{d^2Y}{d\theta^2} + \frac{\cos\theta}{\sin\theta}\frac{dY}{d\theta} + \left(\beta - \frac{m^2}{\sin^2\theta}\right)Y = 0$$

Put $\cos\theta = Z$ then the above equation can be written as

$$-\cos\theta\frac{dY}{dZ} + \sin^2\theta\frac{d^2Y}{dZ^2} + \frac{\cos\theta}{\sin\theta}\left(-\sin\theta\frac{dY}{dZ}\right) + \left(\beta - \frac{m^2}{\sin^2\theta}\right)Y = 0$$

The equation can be simplified as

$$\sin^2\theta\frac{d^2Y}{dZ^2} - 2Z\frac{dY}{d\theta} + \left(\beta - \frac{m^2}{\sin^2\theta}\right)Y = 0$$

Substituting Z for cos θ

$$(1-Z^2)\frac{d^2Y}{dZ^2} - 2Z\frac{dY}{d\theta} + \left(\beta - \frac{m^2}{1-Z^2}\right)Y = 0$$

Which is identical with the differential equation shown in the appendix 8.11 defining the associated Legendre $P_l^m(Z)$ function. Thus we may immediately identify Y(Z) with this function provided that

$$\beta = l(l+1)$$

where $l = 0, 1, 2, 3, \ldots\ldots$

Since

$$E = \frac{h^2}{8\pi^2 I} l(l+1)$$

where $l = 0, 1, 2, 3,\ldots\ldots$

This means that only certain values of E, depending on the value of 'l', are permitted; in other words, the energies of the rigid rotor are quantized.

CHAPTER 5

Applications of Schrödinger Equation-2
(Simple Systems with Variable Potential Energy)

5.1 One-dimensional Harmonic Oscillator

When two atoms, held together firmly in a stable molecule are caused to vibrate, the vibrations may be treated approximately as motions of particles in a harmonic field. It had long been realized that the vibrational spectra of both polyatomic and diatomic molecules should be treated by the theory of a system of particles all moving in a harmonic field. One of the problems of the old quantum theory was to explain the residual energy at 0 K within any system of oscillating particles, the so-called ZERO POINT ENERGY. It was one of the triumphs of the new wave mechanics (and also incidentally of Heisenberg's matrix mechanics) that this residual energy came naturally from the application of the basic postulates of quantum mechanics, in particular that ψ be a well behaved function.

Consider the case of a particle of mass, 'm' moving in one dimension (the 'x' coordinate) in a potential,

$$V = \frac{1}{2} Kx^2 \qquad (5.1.1)$$

and subject to a force $-Kx$. The constant K is called the force constant of the system.

Case 1: In classical mechanics

Classically the equation of motion is

$$m \frac{d^2 x}{dt^2} = -Kx \qquad (5.1.2)$$

with the general solution

$$x = x_0 \cos 2\pi \, v_0 (t - t_0) \qquad (5.1.3)$$

where x_0 and t_0 are constants and

$$v_0 = \frac{1}{2\pi} \left(\frac{k}{m} \right)^{1/2} \qquad (5.1.4)$$

The total energy

$$H = T + V$$
$$= \frac{1}{2m}P_x^2 + \frac{1}{2}Kx^2$$
$$= 2m\pi^2 v_0^2 x_0^2 \qquad (5.1.5)$$

and so all positive values of the energy, including zero, are allowed on the classical picture.

Case 2: In quantum mechanics

In the quantum mechanical case we must again solve the wave equation $H\psi = E\psi$, subject to the condition that ψ is a well behaved function, where we obtain the Hamiltonian operator, H from the classical expression by replacing P_x by $+(h/2\pi i)(d/dx)$ (the momentum operator)

The Hamiltonian operator is thus

$$H = \frac{-h^2}{8\pi^2 m}\frac{d^2}{dx^2} + \frac{Kx^2}{2} \qquad (5.1.6)$$

and so the wave equation becomes

$$\frac{d^2\psi}{dx^2} + \frac{8\pi^2 m}{h^2}\left(E - \frac{Kx^2}{2}\right)\psi = 0 \qquad (5.1.7)$$

We must solve this equation subject to the conditions that ψ be well behaved. The above equation can be written in the more convenient form

$$\frac{d^2\psi}{dx^2} + (\alpha - \beta^2 x^2)\psi = 0 \qquad (5.1.8)$$

Where, $\quad \alpha = \frac{8\pi^2 m}{h^2}E \text{ and } \beta = \frac{2\pi(mh)^{1/2}}{h}$

we require that ψ be well behaved in the complete range of the coordinate x, that is from $-\infty$ to $+\infty$. Let us first consider the asymptotic solution of this equation, that is, the solution for large values of x. In this case 'α' can be neglected when compared with $\beta^2 x^2$ and Eq.5.1.8 becomes

$$\frac{d^2\psi}{dx^2} + \beta^2 x^2 \psi \qquad (5.1.9)$$

which is satisfied by the exponential functions

$$\psi = N.\exp\left(\pm\frac{1}{2}\beta x^2\right) \qquad (5.1.10)$$

where N is a constant.

[Note in checking this, that the second term in the expression for $d_2\psi/dx^2$ is neglected with respect to the first].

However, the solution

$$\psi = N\left[\exp\left(+\frac{1}{2}\beta x^2\right)\right] \quad (5.1.11)$$

is clearly not a well behaved function, since, as x tends to infinity, ψ also tends to infinity and does not remain finite, so only the other solution is acceptable.

It is now convenient to change the variable such that

$$y = x(\beta)^{1/2} \quad (5.1.12)$$

Where

$$\psi(y) = N\exp\left(-\frac{1}{2}y^2\right) \quad (5.1.13)$$

Returning to the general equation, the above asymptotic solution suggests as a possible general solution

$$\psi(y) = f(y)\exp\left(-\frac{1}{2}y^2\right) \quad (5.1.14)$$

where f(y) is a polynomial in y,

Substituting in equation (5.1.8) we find

$$\frac{d^2 f}{dy^2} - 2y\frac{df}{dy} = \left(\frac{\alpha}{\beta} - 1\right)f = 0 \quad (5.1.15)$$

(For details refer Appendix 8.12)

This equation is identical with the Hermite's equation (see appendix 8.11) in which

$$\frac{\alpha}{\beta} - 1 = 2n \quad (5.1.16)$$

In other words, the polynomials Nf (y) are the Hermite polynomials $H_n(y)$.

The wave function

$$\psi(y) = N \quad H_n(y)\exp\left(-\frac{1}{2}y^2\right) \quad (5.1.17)$$

is thus a well behaved function for all positive integral values of 'n' including zero, since it is only for this range that the Hermite polynomials are defined. Eq.5.1.16 which arises in order to terminate the polynomials, so that the product H_n (y) exp(-1/2y_2) tends to zero as y tends to ∞, thus leads to a quantum condition.

$$\frac{\alpha}{\beta} = 2n+1$$

and on substituting for α and β we find

$$E = \frac{h}{2\pi}\left(n+\frac{1}{2}\right)\left(\frac{1}{m}\right)^{1/2} = \left(n+\frac{1}{2}\right)hv_0 \tag{5.1.18}$$

In contrast to the classical result of a continuous spectrum of positive energies, wave mechanics predicts a set of quantized levels whose energies are 1/2, 3/2, 5/2, ... multiplied by the energy of the classical frequency. Moreover wave mechanics shows that the harmonic oscillator cannot have zero energy but always possesses a residual or 'Zero point energy' (1/2) hv_0. This result was a triumph for the new wave mechanics since the older quantum mechanics had not been able to account for this residual energy.

In fact the prediction of zero-point energy by wave mechanics could be linked to the results of matrix mechanics by means of Heisenberg's uncertainty principle, namely that is only possible to specify simultaneously the momentum 'p' and the position 'q' of a particle with uncertainties, Δp and Δq respectively, where $\Delta p \cdot \Delta q \approx h$.

Wave functions of the harmonic oscillator

The wave functions for the harmonic oscillator were shown to be

$$\psi(y) = N \quad H_n(y)\exp\left(-\frac{1}{2}y^2\right)$$

where $H_n(y)$ are Hermite polynomials and $y = x(\beta)^{\frac{1}{2}}$

The normalization constant N is obtained from the usual condition

$$\int_{-\infty}^{+\infty} \psi_n^* \psi_n \, d\tau = 1$$

from which it can be shown that

$$N = \left[\left(\frac{\beta}{\pi}\right)^{1/2} \frac{1}{2^n N}\right]^{1/2}$$

and so the normalized wave functions for the one-dimensional harmonic oscillator are

$$\psi_n(y) = \left[\left(\frac{\beta}{\pi}\right)^{1/2} \frac{1}{2^n N}\right]^{1/2} \left[H_n(y)\exp\left(-\frac{1}{2}y^2\right)\right]$$

where $y = x(\beta)^{\frac{1}{2}}$. It is instructive to list a few of the earlier members of the Hermite polynomials:

$H_0(y) = 1$ $H_3(y) = 8y_3 - 12y$
$H_1(y) = 2y$ $H_4(y) = 16y_4 - 48y_2 + 12$
$H_2(y) = 4y_2 - 2$ $H_5(y) = 32y_5 - 160y_3 + 120y$

It is also interesting to examine the form of the above wave functions for increasing values of 'n', the vibrational quantum number. The case of n = 0 is the most interesting, since, in this case both $\psi_0(y)$ and $\psi_0^2(y)$ (where $\psi_0^2(y)dy$ measures the probability of finding the system in this state between y and y + dy) possesses maxima at the origin, or zero probability at the extreme of the vibrational mode. This is the exact opposite of classical mechanics, where one would expect the particle to be most probably at the ends of its motion. As discussed above, this contrast is due to the quantum mechanical result of the particle possessing zero point energy, which in turn can be correlated with the Heisenberg's uncertainty principle. As 'n' increases we note that the wave functions are alternatively symmetric and antisymmetric and finally as 'n' becomes very large (see Fig. 5.1.1b) the probability function $\psi_0^2(y)$ approximates to the classical probability curve (dotted line).

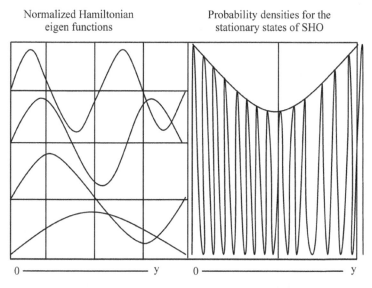

Fig. 5.1.1a Wave functions of harmonic $\psi_n(y)$ plotted against y.

Fig. 5.1.1b The square of the wave functions of a harmonic oscillator $\psi_n^2(y)$ plotted against y.

According to correspondence principle, at high quantum numbers, the oscillator should approach a macroscopic classical oscillator in behavior, e.g., a pendulum. The figure shows that the probability distribution is concentrated near the "turning points" where the system's energy is purely potential. This indeed describes a pendulum, which spends more time slowing down, reversing direction, and accelerating near the extremes of displacement than in the centre of the movement where it has its maximum speed.

5.2 The Hydrogen Atom

Transformation of coordinates: A new problem arises when we note that the total energy, E in the wave function is made up of two parts as in the case of hydrogen atom.

1. The translational motion of the atom as a whole.
2. The energy of the electron with respect to the proton.

It is this latter portion of the energy, in which we are interested. This leads us again to the problem of separation of variables. In order to obtain the desired equation, it will be necessary to separate and discard the translational portion of the total wave equation. To carry out this particular separation, it is necessary to introduce a new set of variables x, y and z, which are Cartesian coordinates of the centre of mass of the hydrogen atom and the variables r, θ and φ which are Poplar coordinates of the electron with respect to the nucleus.

A coordinate of the centre of mass of a system in general, given by

$$q = \frac{\sum_i m_i q_i}{\sum_i m_i}$$

where, m_i is the centre of mass of the 'i'th particle. For hydrogen atom, the Cartesian coordinates of the centre of mass will be given by

$$x = \frac{m_1 x_1 + m_2 x_2}{m_1 + m_2}$$

$$y = \frac{m_1 y_1 + m_2 y_2}{m_1 + m_2}$$

$$z = \frac{m_1 z_1 + m_2 z_2}{m_1 + m_2}$$

But for our purpose the transformation into spherical coordinates is important.

Polar coordinates

The problem we are concerned with is that of calculating the "amplitude of the electron waves" at various points in a hydrogen atom. These points can be defined by drawing a set of Cartesian (x, y, z) axes through the origin at the nucleus of the atom and locating points on x, y and z coordinates. It is much simpler, if we use an alternate way of specifying position, namely the polar coordinate system. These are shown in the following figure.

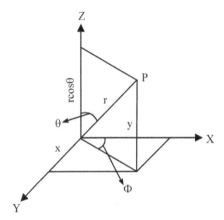

Fig. 5.2.1 Transformation of Cartesian coordinates into Polar coordinates.

$$p = r \sin \theta$$

$$\cos \phi = \frac{x}{s}, \qquad \therefore x = r \sin \theta . \cos \phi$$

$$\sin \phi = \frac{y}{s} \qquad \therefore y = r \sin \theta . \sin \phi$$

$$\cos \theta = \frac{z}{r} \qquad \therefore z = r \cos \theta$$

$$x^2 + y^2 + z^2 = r^2$$

These transformations into spherical coordinates in general is written as

$$x_2 - x_1 = r \sin \theta . \cos \phi$$

$$y_2 - y_1 = r \sin \theta . \sin \phi$$

$$z_2 - z_1 = r \cos \theta$$

By using the transformation equations, it is a straightforward procedure to obtain the wave equation in terms of Cartesian coordinates of the centre of mass of the system and the polar coordinates r, θ and φ. The x, y, z coordinates of the centre of mass of the atom obviously relate to the translational motion of the atom as a whole and r, θ and φ coordinates are seen to relate the coordinates of the electron (x_1, y_1, z_1) to the coordinates of the nucleus (x_2, y_2, z_2).

As an example of the procedure, consider the z coordinate. Solving the equation for z

$$z = \frac{m_1 z_1 + m_2 z_2}{m_1 + m_2}$$

$$z(m_1 + m_2) = m_1 z_1 + m_2 z_2$$

$$\therefore z_2 = z\left(\frac{m_1 + m_2}{m_2}\right) - \frac{m_1}{m_2} z_1$$

If the value of z_2 is substituted in the equation $z_2 - z_1 = r \cos\theta$

$$r\cos\theta = \left(\frac{m_1 + m_2}{m_2}\right) - \frac{m_1}{m_2} z_1 - z_1$$

$$z_1\left(\frac{m_1}{m_2} + 1\right) + r\cos\theta = z\left(\frac{m_1 + m_2}{m_2}\right)$$

$$z_1\left(\frac{m_1 + m_2}{m_2}\right) = z\left(\frac{m_1 + m_2}{m_2}\right) - r\cos\theta$$

Dividing throughout by $\frac{m_1 + m_2}{m_2}$

$$z_1 = z - \frac{m_2}{m_1 + m_2} r\cos\theta$$

$$z_1 = z - \left(\frac{m_2}{m_1 + m_2}\right)\frac{m_1}{m_1} r\cos\theta$$

$$z_1 = z - \frac{\mu}{m_1} r\cos\theta \text{ where } \mu = \frac{m_1 m_2}{m_1 + m_2}$$

Using the procedure, a transformation equation can be found for each of the coordinates, and when proper substitutions are made, the wave equation is obtained in terms of the variables x, y, z, r, θ and φ. In terms of the new variables, it is found to be

Chapter 5 | Applications of Schrödinger Equation-2

$$\frac{1}{m_1 + m_2}\left(\frac{\partial^2 \psi_T}{\partial x^2} + \frac{\partial^2 \psi_T}{\partial y^2} + \frac{\partial^2 \psi_T}{\partial z^2}\right) +$$

$$\frac{m_1 + m_2}{m_1 m_2}\left[\frac{1}{r^2}\frac{\partial}{\partial r}\left(r^2 \frac{\partial \psi_T}{\partial r}\right) + \frac{1}{r^2 \sin^2 \theta}\frac{\partial \psi_T}{\partial \phi^2} + \frac{1}{r^2 \sin \theta}\frac{\partial}{\partial \theta}\left(\sin \theta \frac{\partial \psi_T}{\partial \phi^2}\right)\right]$$

$$+ \frac{8\pi^2}{h^2}(E - V_r)\psi_T = 0 \qquad (5.2.1)$$

The wave equation ψ_{Total} is a function of the variables x, y, z, r, θ, ϕ and the energy, E contains the translational energy as well as the energy of the electron with respect to the proton.

The purpose of this transformation to new coordinates is to make a separation of variables possible. The algebra is somewhat more complex. In the usual manner, the total wave function, $\psi_{(x,y,z,r,\theta,\phi)}$ is assumed to be expressible as the product of two wave functions such that

$$\psi_{(x,y,z,r,\theta,\phi)} = F_{(x,y,z)}\psi_{(r,\theta,\phi)}$$

When this expression is substituted into Eq. 5.2.1, it is found that the following two equations are obtained.

$$\frac{\partial^2 F_{(x,y,z)}}{\partial x^2} + \frac{\partial^2 F_{(x,y,z)}}{\partial y^2} + \frac{\partial^2 F_{(x,y,z)}}{\partial z^2} + \frac{8\pi^2(m_1 + m_2)}{h^2}E_{trans}F_{(x,y,z)} = 0$$

$$\frac{1}{r^2}\frac{\partial}{\partial r}\left(r^2 \frac{\partial \psi_{(r,\theta,\phi)}}{\partial r}\right) + \frac{1}{r^2 \sin^2 \theta}\frac{\partial^2 \psi_{(r,\theta,\phi)}}{\partial \phi^2} + \frac{1}{r^2 \sin \theta}\frac{\partial}{\partial \theta}\left(\sin \theta \frac{\partial \psi_{(r,\theta,\phi)}}{\partial \phi}\right) +$$

$$\frac{8\pi^2 \mu}{h^2}(E - V_r)\psi_{(r,\theta,\phi)} = 0$$

The first of these equations contains only the variable x, y and z with no potential energy term. This is identical to the wave equation for a free particle and therefore represents the translational energy as a whole. The second equation, which relates the electron to the proton, is the equation of particular interest to us.

Separation of variables

Since it is the second part of the wave equation that is of interest, the translational part will be discarded. This is the desired equation for the electron with respect to nucleus. This equation contains three variables r, θ and ϕ. This will require that the variables be separated, such that, three independent equations are obtained, each containing only one of the three variables.

Quantum Chemistry

The procedure is as follows. The Schrödinger equation in polar coordinates can be written as:

$$\frac{1}{r^2}\frac{\partial}{\partial r}\left(r^2 \frac{\partial \Psi_{(r\theta\phi)}}{\partial r}\right) + \frac{1}{r^2 \sin^2\theta}\frac{\partial^2 \Psi_{(r\theta\phi)}}{\partial \phi^2} + \frac{1}{r^2 \sin\theta}\frac{\partial}{\partial \theta}\left(\sin\theta \frac{\partial \Psi_{(r\theta\phi)}}{\partial \phi}\right) +$$

$$\frac{8\pi^2 \mu}{h^2}(E - V_r)\Psi_{(r\theta\phi)} = 0 \qquad (5.2.2)$$

Let $\Psi_{(r,\theta,\phi)} = R_{(r)}\Theta_{(\theta)}\Phi_{(\phi)}$

Making substitution in the above equation

$$\frac{1}{r^2}\frac{\partial}{\partial r}\left(r^2 \frac{\partial R_{(r)}\Theta_{(\theta)}\Phi_{(\phi)}}{\partial r}\right) + \frac{1}{r^2 \sin^2\theta}\frac{\partial^2 R_{(r)}\Theta_{(\theta)}\Phi_{(\phi)}}{\partial \phi^2} +$$

$$\frac{1}{r^2 \sin\theta}\frac{\partial}{\partial \theta}\left(\sin\theta \frac{\partial R_{(r)}\Theta_{(\theta)}\Phi_{(\phi)}}{\partial r}\right) + \frac{8\pi^2\mu}{h^2}(E-V_r)R_{(r)}\Theta_{(\theta)}\Phi_{(\phi)} = 0 \qquad (5.2.3)$$

Dividing Eq.5.2.3 by $R_{(r)}\Theta_{(\theta)}\Phi_{(\phi)}$ we obtain

$$\frac{1}{r^2 R(r)}\frac{\partial}{\partial r}\left(r^2 \frac{\partial R_{(r)}}{\partial r}\right) + \frac{1}{\Phi_{(\phi)}r^2 \sin^2\theta}\frac{\partial^2 \Phi_{(\phi)}}{\partial \phi^2} + \frac{1}{\Theta_{(\theta)}r^2 \sin\theta}\frac{\partial}{\partial \theta}\left(\sin\theta \frac{\partial \Psi_{(\theta)}}{\partial \theta}\right) +$$

$$\frac{8\pi^2\mu}{h^2}(E-V_r) = 0$$

If we multiply by $r^2 \sin^2\theta$

$$\frac{\sin^2\theta}{R_{(r)}}\frac{\partial}{\partial r}\left(r^2 \frac{\partial R_{(r)}}{\partial r}\right) + \frac{1}{\Phi_{(\phi)}}\frac{\partial^2 \Phi_{(\phi)}}{\partial \phi^2} + \frac{\sin\theta}{\Theta_{(\theta)}}\frac{\partial}{\partial \theta}\left(\sin\theta \frac{\partial \Theta_{(\theta)}}{\partial \theta}\right) +$$

$$\frac{8\pi^2 \mu r^2 \sin^2\theta}{h^2}(E-V_r) = 0$$

or

$$\frac{\sin^2\theta}{R_{(r)}}\frac{\partial}{\partial r}\left(r^2 \frac{\partial R_{(r)}}{\partial r}\right) + \frac{\sin\theta}{\Theta_{(\theta)}}\frac{\partial}{\partial \theta}\left(\sin\theta \frac{\partial \Theta_{(\theta)}}{\partial \theta}\right) +$$

$$\frac{8\pi^2 \mu r^2 \sin^2\theta}{h^2}(E-V_r) = -\frac{1}{\Phi_{(\phi)}}\frac{\partial^2 \Phi_{(\phi)}}{\partial \phi^2} \qquad (5.2.4)$$

The left side of the equation contains only the variables r and θ where the right side of the equation contains only the variable ϕ. No matter what values r

Chapter 5 | Applications of Schrödinger Equation-2

and θ might independently take the sum of the terms on the left always equal to the term on the right. This can be true only if each side of the equation is equal to the same constant. If we let this constant be 'm²', it is seen that the variable can be immediately be separated from Eq.5.2.4 giving

$$\frac{1}{\phi_{(\phi)}}\frac{\partial^2 \phi_{(\phi)}}{\partial \phi^2} = -m^2 \tag{5.2.5}$$

The problem now is to carry out the separation of the remaining two variables r and θ. By equating the LHS of Eq. 5.2.4 to m², it is seen that

$$\frac{\sin^2\theta}{R_{(r)}}\frac{\partial}{\partial r}\left(r^2 \frac{\partial R_{(r)}}{\partial r}\right) + \frac{\sin\theta}{\theta_{(\theta)}}\frac{\partial}{\partial \theta}\left(\sin\theta \frac{\partial \theta_{(\theta)}}{\partial \theta}\right) + \frac{8\pi^2 \mu r^2 \sin^2\theta}{h^2}(E - V_r) = m^2$$

On division by $\sin^2\theta$, this becomes

$$\frac{1}{R_{(r)}}\frac{\partial}{\partial r}\left(r^2 \frac{\partial R_{(r)}}{\partial r}\right) + \frac{8\pi^2 \mu r^2}{h^2}(E - V_r) = \frac{m^2}{\sin^2\theta} - \frac{1}{\theta_{(\theta)}\sin\theta}\frac{\partial}{\partial \theta}\left(\sin\theta \frac{\partial \theta_{(\theta)}}{\partial \theta}\right)$$

Again since each side of the equation contains only one variable, they both must be equal to the same constant. If the right side of the equation is set equal to the constant β and this gives on multiplication $\theta_{(\theta)}$ by

$$\frac{m^2}{\sin^2\theta} - \frac{1}{\theta_{(\theta)}\sin\theta}\frac{\partial}{\partial \theta}\left(\sin\theta \frac{\partial \theta_{(\theta)}}{\partial \theta}\right) - \beta\theta_{(\theta)} = 0 \tag{5.2.6}$$

This is the desired form of the θ equation.

The remaining part of the original equation is the R equation.

$$\frac{1}{R_{(r)}}\frac{\partial}{\partial r}\left(r^2 \frac{\partial R_{(r)}}{\partial r}\right) + \frac{8\pi^2 \mu r^2}{h^2}(E - V_r) = \beta$$

This equation on multiplication throughout with $\frac{R_{(r)}}{r^2}$ and rearranging gives

$$\frac{1}{r^2}\frac{\partial}{\partial r}\left(r^2 \frac{\partial R_{(r)}}{\partial r}\right) - \beta\frac{R_{(r)}}{r^2} + \frac{8\pi^2 \mu r^2}{h^2}(E - V_r) = R_{(r)} = 0 \tag{5.2.7}$$

Thus, the three variables have been successfully separated, and the three independent total differential equations that result are:

$$\frac{1}{\phi_{(\phi)}}\frac{\partial^2 \phi_{(\phi)}}{\partial \phi^2} = -m^2 \tag{5.2.5}$$

$$\frac{m^2}{\sin^2\theta} - \frac{1}{\theta_{(\theta)}\sin\theta}\frac{\partial}{\partial\theta}\left(\sin\theta\frac{\partial\theta_{(\theta)}}{\partial\theta}\right) - \beta\theta_{(\theta)} = 0 \qquad (5.2.6)$$

$$\frac{1}{r^2}\frac{\partial}{\partial r}\left(r^2\frac{\partial R_{(r)}}{\partial r}\right) - \beta\frac{R_{(r)}}{r^2} + \frac{8\pi^2\mu r^2}{h^2}(E - V_r) = R_{(r)} = 0 \qquad (5.2.7)$$

The ϕ equation

The first of these equations is the ϕ equation and it is seen to be of the same form as the wave equation for the particle in a box.

In terms of the sine and cosine the solution is

$$\phi_{m(\phi)} = A\sin m\phi + B\cos m\phi$$

In order for a wave function to be acceptable, it must be of the well behaved class. One of the requirements of such a function is that it must be single valued. To meet this restriction, the function $\phi_{m(\phi)}$ must have the same value for $\phi = 0$ as it does for $\phi = 2\pi$. For the case of $\phi = 0$ it can be seen that

$$\phi_{m(0)} = A\sin 0 + B\cos 0$$

$$\phi_{m(0)} = B$$

When we have $\phi = 2\pi$,

$$\phi_{m(2\pi)} = A\sin m2\pi + B\cos m2\pi$$

$$\therefore \phi_{m(2\pi)} = B.$$

Since the value of ϕ must be the same under both the conditions, it is necessary that

$$B = A\sin m2\pi + B\cos m2\pi.$$

This identity can hold only if 'm' is zero or has a positive or negative integral value. So the possible values of 'm' are 0, ±1, ±2, ±3, …… and 'm' is known as the *magnetic quantum number*.

Very often, in the treatment of hydrogen atom, the exponential solution to the ϕ equation

$$\phi_{m(\phi)} = ce^{\pm im\phi}$$

is used. This is shown as follows:

The equation is

$$\frac{\partial^2\phi_{(\phi)}}{\partial\phi^2} + m^2\phi_{(\phi)} = 0 \qquad (5.2.5)$$

The wave function $\phi_{m(\phi)}$ is to be normalized.

This requires that $\int_0^{2\pi} \phi\phi^* d\phi = 1$

Which leads to $\int_0^{2\pi} c^2 e^{im\phi} e^{-im\phi} d\phi = 1$

$$c^2 \int_0^{2\pi} e^0 d\phi = 1 \text{ or } [\phi]_0^{2\pi} = \frac{1}{c^2}$$

$$\therefore \quad c = \frac{1}{\sqrt{2\pi}}.$$

Hence the normalized function is

$$\phi_{m(\phi)} = \frac{1}{\sqrt{2\pi}} c e^{\pm im\phi}.$$

This is also a solution to the wave equation and is true only if 'm' can have integral values starting from $0, \pm1, \pm2, \pm3, \ldots\ldots$

The θ equation

The equation is

$$\frac{m^2}{\sin^2\theta} - \frac{1}{\theta_{(\theta)} \sin\theta} \frac{\partial}{\partial\theta}\left(\sin\theta \frac{\partial\theta_{(\theta)}}{\partial\theta}\right) - \beta\theta_{(\theta)} = 0 \quad (5.2.6)$$

This equation can be put into a form that was known by the mathematicians many years before the advent of quantum mechanics. This particular equation is known as Legendre's equation and has the normalized solution

$$\theta_{l,m(\theta)} = \sqrt{\frac{(2l+1)(l-|m|)!}{2(l+|m|)!}} P_l^{|m|}(\cos\theta)$$

where $P_l^{|m|}$ is the associated Legendre function of degree 'l' and order $|m|$. The form of the solution is quite complicated. In spite of the complicated nature of the solution, several important features can be observed. Although the mathematics is far too complex to be considered here, it can be shown that in Eq. (3.2.5), $\beta = l(l+1)$, where the allowed values of l are 0, 1, 2, 3….. This is the source of the new parameter found in Eq.(3.2.7) and its properties appear to be similar in many ways to those of the *azimuthal quantum number*. It can also be seen that there is now a new restriction on the quantum number 'm'. In the normalizing factor of the solution to the θ equation, the term $(l-|m|)!$ occurs. If $|m|$ is allowed to be greater than 'l', the factorial of a negative number results. Since a negative factorial is undefined, the maximum value of 'm' must be 'l'. Thus the restrictions on the quantum number 'm' now become m = 0, ±1, ±2, ±3, ……±l.

Spherical Harmonics

Both the solution to the θ equation and solution to the ϕ equation contain trigonometric functions and therefore determine the angular character of the electron wave function. Very often it is found that the total wave function can most conveniently be used, if it is separated into a radial portion and an angular portion such that

$$\psi_{(r\theta\phi)} = R_{n,l(r)} Y_{lm(\phi)} \qquad (5.2.8)$$

The term is referred to as the spherical harmonics and is given by

$$Y_{lm(\theta\phi)} = \Theta_{l,m(\theta)} \Phi_{m(\phi)} \qquad (5.2.9)$$

This portion of the wave function is important in the treatment of directional bonding.

The Radial Equation

The remaining equation to be solved is the radial equation.

$$\frac{1}{r^2}\frac{\partial}{\partial r}\left(r^2 \frac{\partial R_{(r)}}{\partial r}\right) - \beta \frac{R_{(r)}}{r^2} + \frac{8\pi^2 \mu r^2}{h^2}\left(E - V_{(r)}\right) R_{(r)} = 0 \qquad (5.2.7)$$

This like the θ equation, can be put into a form that has long been known to mathematicians. This particular equation is the Laguerre equation, and the normalized solution is

$$R_{n,l(r)} = \sqrt{\left(\frac{2z}{na_0}\right)^3 \frac{(n-l-1)!}{2n([n+1]!)^3}} \; e^{-\frac{\rho}{2}} \rho^l L_{n+l}^{2l+1} \qquad (5.2.10)$$

$$\rho = \frac{2z}{na_0} r,$$

where $\quad a_0 = \dfrac{h^2}{4\pi^2 \mu e^2}$

and $L_{n+l}^{2l+1}(\rho)$ represents the associated Laguerre polynomial.

Chapter 5 | Applications of Schrödinger Equation-2

The solution to the radial equation is also very complex. However, it is possible to make pertinent observations from the solution. It is to be noted that a new parameter, the quantum number 'n', has been added. The possibility here is that 'n' is restricted to take on only the integral values 1, 2, 3, ... Both the relation of 'n' to the radial wave function, which is a measure of the position of the electron with respect to the nucleus, and its similar restrictions, indicate that 'n' is the quantum mechanical analog of the principal quantum number.

A new restriction can be seen for quantum number 'l'. It is apparent that the term $(n-l-1)!$ requires that the maximum value of l be $(n-1)$. If 'l' is allowed a value greater than this, the factorial of a negative number would result. Since a negative factorial is undefined, the maximum value of 'l' be $(n-1)$. Thus the quantum number is restricted to the values $l = 0, 1, 2, 3, \ldots, (n-1)$.

Quantum States

From the solution of the total wave equation, we have arrived at three quantum numbers. The quantum numbers with their allowed values may be summarized as follows:

Radial quantum number $n = 1, 2, 3, \ldots$

Azimuthal quantum number $l = 0, 1, 2, \ldots, (n-1)$

Magnetic quantum number $m = 0, \pm 1, \pm 2, \pm 3, \ldots \pm$

According to these restrictions, there are only certain values of the quantum number, 'l' that are permissible, for a given value of 'n'. The maximum value of 'l' is seen always to be $(n-1)$. For example, when 'n' = 4, 'l' can be any integer up to and including 3, but not greater than that. This is illustrated as follows:

Value of 'n'	1	2	3	4
Allowed value of 'l'	0	0,1	0, 1, 2	0, 1, 2, 3

It should be noted that $l = 0$ occurs for every value of 'n'; $l = 1$ occurs for every value of 'n' greater than n = 1, and so on.

These values of quantum number l, play a rather important role in both the geometry and energy states of the atom. Because of this importance, they are given the following special designations.

     $l = 0$ s state

     $l = 1$ p state

     $l = 2$ d state

     $l = 3$ f state

Quantum Chemistry

For the first radial shell, the value of the radial quantum number is $n = 1$, and l quantum number can only have the value $l = 0$. This state is usually represented by (1s) where '1' represents the principal quantum number. For $n = 2$, the azimuthal quantum number can have the values $l = 0$ and $l = 1$. This gives the two states (2s) and (2p) respectively. These states determine the energies of electrons, and if the l quantum number contributes to the energy as does the 'n' quantum number, each state will represent a different energy.

Wave Functions of the Hydrogen Atom

It was postulated that the square of the wave function is a measure of the probability distribution of the electron. This wave function is seen to be composed of two parts, an angular portion represented by $Y_{l,m(\theta,\phi)}$ and a radial portion that is represented by $R_{n,l(r)}$. We will see that the radial portion of the wave function gives the distribution of the electron with respect to the distance from the nucleus, whereas the angular portion gives the geometry of the various energy states.

The normalized solutions of the θ equation, and also the radial equation are in general, quite complex. However, they reduce to relatively simple form on introduction of particular values of the parameters. For the 'ϕ' equation the allowed values of 'm' are $m = 0, \pm 1, \pm 2, \pm 3, ..\pm l$. This led to the normalized functions of shown below.

Normalized functions of $\phi_{m(\phi)}$

$$\phi_{m(\phi)} = \frac{1}{\sqrt{2\pi}} e^{\pm im\phi}$$

$$\phi_{0(\phi)} = \frac{1}{\sqrt{2\pi}} e^{0} \text{ or } \phi_{0(\phi)} = \frac{1}{\sqrt{2\pi}}$$

$$\phi_{1(\phi)} = \frac{1}{\sqrt{2\pi}} e^{\pm i\phi} \text{ or } \phi_{1(\phi)} = \frac{1}{\sqrt{\pi}}\cos\phi$$

$$\phi_{-1(\phi)} = \frac{1}{\sqrt{2\pi}} e^{-i\phi} \text{ or } \phi_{-1(\phi)} = \frac{1}{\sqrt{\pi}}\sin\phi$$

Examples of the normalized $\theta_{(\theta)}$ functions and the radial functions are given below.

Normalized functions of $\theta_{l,m(\theta)}$.

$$\theta_{l,m(\theta)} = \sqrt{\frac{(2l+1)(l-|m|)!}{2(l-|m|)!}}\, P_l^{|m|}(\cos\theta)$$

Chapter 5 | Applications of Schrödinger Equation-2

$l = 0 \qquad \theta_{0,0(\theta)} = \dfrac{2}{\sqrt{2}}$

$l = 1 \qquad \theta_{1,0(\theta)} = \dfrac{\sqrt{6}}{2}\cos\theta$

$\qquad\qquad \theta_{1,\pm 1(\theta)} = \dfrac{\sqrt{3}}{2}\sin\theta$

Normalized functions of $R_{n,l(r)}$

$$R_{n,l(r)} = \sqrt{\left(\dfrac{2z}{na_0}\right)^3 \dfrac{(n-1-1)!}{2n\left[(n+l)!\right]^3}}\; e^{-\frac{\rho}{2}}\rho^{l} L_{n+1}^{2l+1}$$

n = 1, 'K' shell

$l = 0 \quad R_{1,0(r)} = \left(\dfrac{2}{a_0}\right)^{\frac{3}{2}}.2e^{-\frac{\rho}{2}}$

n = 2, 'L' shell

$l = 0 \quad R_{2,0(r)} = \dfrac{\left(\dfrac{z}{a_0}\right)^{\frac{3}{2}}}{2\sqrt{2}}.(2-\rho)e^{-\frac{\rho}{2}}$

$l = 1 \quad R_{2,1(r)} = \dfrac{\left(\dfrac{z}{a_0}\right)^{\frac{3}{2}}}{2\sqrt{6}}.\rho e^{-\frac{\rho}{2}}$

The normalized total wave function for the hydrogen atom is obtained from relation

$$\psi_{(r\phi\theta)} = R_{n,l(r)} Y_{lm(\theta\phi)}.$$

By using all possible arrangements of these functions within the limitations of the quantum numbers, we obtain the normalized total wave functions listed below.

K shell

$n = 1, l = 0, m = 0; \quad \psi_{1s} = \dfrac{1}{\sqrt{\pi}}\left(\dfrac{z}{a_0}\right)^{\frac{3}{2}} e^{-\varepsilon}$

L shell

$$n = 2, l = 0, m = 0; \quad \psi_{2s} = \frac{1}{4\sqrt{2\pi}} \left(\frac{z}{a_0}\right)^{\frac{3}{2}} (2-\varepsilon) e^{-\frac{\varepsilon}{2}}$$

$$n = 2, l = 1, m = 0; \quad \psi_{2p_z} = \frac{1}{4\sqrt{2\pi}} \left(\frac{z}{a_0}\right)^{\frac{3}{2}} \varepsilon e^{-\frac{\varepsilon}{2}} \cos\theta$$

$$n = 2, l = 1, m = \pm 1;$$

$$\psi_{2p_x} = \frac{1}{4\sqrt{2\pi}} \left(\frac{z}{a_0}\right)^{\frac{3}{2}} \varepsilon e^{-\frac{\varepsilon}{2}} \sin\theta \cos\phi$$

$$\psi_{2p_y} = \frac{1}{4\sqrt{2\pi}} \left(\frac{z}{a_0}\right)^{\frac{3}{2}} \varepsilon e^{-\frac{\varepsilon}{2}} \sin\theta \sin\phi$$

Hydrogen like Wave Functions

The Radial Function

These functions are independent of θ and ϕ. They are spherically symmetrical, so that we get the same value of $R_{(r)}$ at a given distance 'r' from the nucleus no matter what values are given to θ and ϕ.

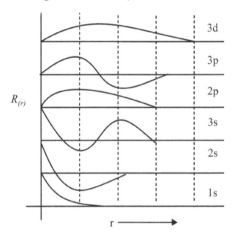

Fig. 5.2.2 Plots of radial function $R_{(r)}$ against r.

It is seen that

(a) All the 's' functions have their maximum values at the nucleus ($r = 0$) and that the value of $R_{(r)}$ initially drops very steeply as 'r' increases. This is because the mathematical expression for each solution includes a negative exponential expression of the type e^{-r}.

Functions of this type decrease rapidly from a maximum value at $r = 0$ to zero at $r = \infty$.

(b) $R_{(r)}$ for the 2s orbital become zero at a particular value of 'r' between $r = 0$ and $r = \infty$. At this point, the so called nodal point, changes sign from positive to negative. The 3s orbital has two nodal points between zero and ∞. In general, the number of such nodes in a 'ns' orbital is given by $(n-1)$.

(c) The p and d radial functions are all zero at $r = 0$ and $r = \infty$ and the number of nodes between these limits is given by $n - l - 1$. Thus there are no nodes in this region in the 2p and 3d radial functions, but there is no node in the 3p function.

(d) At distances close to the nucleus the $R_{(r)}$ function for 's' orbitals is greater than that for p and d orbitals of the same quantum number.

The Radial Distribution Functions

The radial functions $R_{(r)}$ have no physical significance in themselves, but the square of the functions, multiplied by a volume element d_v, $R_{(r)}^2 d_r$, measures the probability that the electron will be in this volume element d_v, at a point that is at a distance 'r' from the nucleus. A more useful value is the probability of finding the electron at a distance 'r' from the nucleus, irrespective of the values of θ and ϕ. Instead of the volume element d_v we now consider the value of a spherical shell of thickness d_r and radius r.

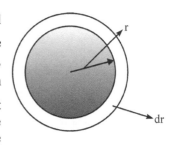

The volume of that will be $4\pi r^2 dr$, since the surface area of the sphere of radius r is $4\pi r^2$. Hence, the radial distribution function $4\pi r^2 R_{(r)}^2 dr$, thus measures the probability of finding the electron in a spherical shell of thickness 'dr' at various distances 'r' from the nucleus.

The following figure 5.2.3. shows plots of this function for the hydrogen 1s, 2s, and 3s orbitals. The functions differ from the simple radial function $R_{(r)}$ in that they are always zero at $r = 0$. The number of peaks the distribution function has for 's' orbitals is equal to the 'n' values i.e. 1 for 1s, 2 for 2s etc.

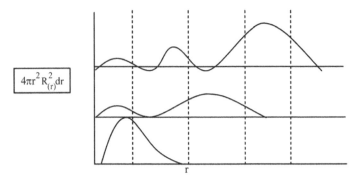

Fig. 5.2.3 Plot of radial distribution function $4\pi r^2 R_{(r)}^2 dr$ against 'r' for 's' electrons.

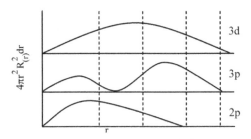

Fig.5.2.4 Plot of radial distribution function $4\pi r^2 R_{(r)}^2 dr$ against 'r' for 'p' and 'd' electrons.

Fig. 5.2.4 shows radial distribution function for 2p, 3p and 3d orbitals. Here the number of peaks is (n-1) for p orbitals and (n-2) for d orbitals.

These functions are particularly useful when discussing the screening effect of electrons in many electron atoms and the peaks of maximum probability corresponds to concentric shells strongly resembling the Bohr theory.

If we go back to the Bohr picture of 1s orbit, we recall that the electron moved in a circular path of fixed distance from the nucleus. If we plot $\psi^2 r^2$ for such a situation, we will get a diagram as shown in Fig. 4a. The probability will be zero at all except the single value when $r = a_0$, where a_0 is the radius of the first orbit (0.529 Å). The corresponding probability function for a 1s electron in the wave mechanical model will be as shown in Fig.4b.

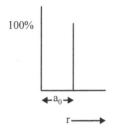
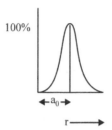

Fig. 5.2.5a Probability function of 1s orbital in classical model

Fig. 5.2.5b Probability function of 1s orbital in wave mechanical model

In both the models electron is found with the highest probability at 0.529 $\overset{o}{A}$ from the nucleus, but in Bohr model this probability comes out to be a certainty (100 % probability). In both the models, the electron density is spherically symmetrical. However, the Bohr model violates the uncertainty principle by fixing the exact radius of the orbit. The wave mechanical model on the other hand "spreads the electrons out" over all space, and so is in accord with Heisenberg's principle.

Show that $r = a_0$ for the 1s orbital

To get the value of 'r', we can take the relevant portion of the radial distribution function. This is differentiated and equated to zero since at this point there is a maximum (Fig 5.2.5b). For 1s orbital in hydrogen ($z = 1$) the wave function is represented as $r^2 e^{\frac{-2r}{a_0}}$.

Hence, radial distribution function = const. $\times r^2 e^{\frac{-2r}{a_0}}$.

According to the rule of maximization, $\dfrac{dR_{(r)}}{dr} = 0$.

On differentiation

$$\frac{dR_{(r)}}{dr} = 0 = 2r.e^{\frac{-2r}{a_0}} - r^2 \frac{2}{a_0} e^{\frac{-2r}{a_0}}$$

or $\quad 2r.e^{\frac{-2r}{a_0}} - r^2 \dfrac{2}{a_0} e^{\frac{-2r}{a_0}}$

or $\quad r = a_0$.

From the above calculations, we can see that the most probable position for the electron is identical with the radius predicted by Bohr for the first electron orbit.

The Angular Function $Y_{(\theta,\phi)}$

We have seen that wave functions for all 's' orbitals are spherically symmetrical i.e. they are independent of the angles θ and ϕ. There are three angular functions for orbitals with n = 2 and l = 1. These are the $2p_0$, $2p_{+1}$ and $2p_{-1}$ orbitals. Similarly there are five angular functions for 3d orbitals corresponding to the five values for the 'm' quantum number where l = 2.

There are two methods.

1. A polar graph is drawn by plotting the θ dependent function against different values of θ for given value of l and m (Fig.5.2.6a).
2. A polar graph is drawn by plotting the square of θ dependent part of the wave function against different values of θ for given values of l and m. This type of graph gives a map of the angular distribution of electron density (Fig.5.2.6b).

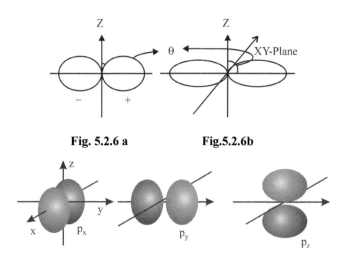

Fig. 5.2.6 a   **Fig.5.2.6b**

Fig. 5.2.7 illustrates polar diagrams for the three 'p' orbitals.

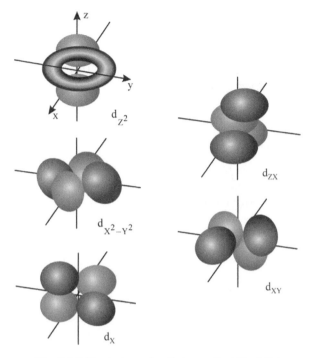

Fig. 5.2.8 illustrates polar diagrams for 'd' orbitals.

Nomenclature of p Orbitals

The nomenclature of p and d orbitals arises from the following relations. If the Cartesian coordinates are replaced by the polar coordinates

1. The functions for ψ_{p_x}, ψ_{p_y} and ψ_{p_z} are

$$\psi_{211} = \frac{1}{4\sqrt{(2\pi a^3)}a} \frac{r}{a} e^{\frac{-r}{2a}} \sin\theta.\cos\phi$$

$$= \text{constant} \times \sin\theta.\cos\phi$$

$$= p_x$$

similarly

$$\psi_{2,1,-1} = \text{constant} \times \sin\theta.\cos\phi$$

$$= p_y$$

and $\quad \psi_{2,1,0} = \text{constant} \times \cos\theta$

$$= p_z.$$

Nomenclature of d Orbitals

The wave functions are

$$\psi_{3,2,1} = \frac{\sqrt{2}}{81\sqrt{(\pi a^3)}} \frac{r^2}{a^2} e^{\frac{-r}{3a}} \sin\theta.\cos\theta.\cos\phi$$

$$= \text{constant} \times \sin\theta.\cos\theta.\cos\phi$$

$$= d_{xz}.$$

$$\psi_{3,2,-1} = \frac{\sqrt{2}}{81\sqrt{(\pi a^3)}} \frac{r^2}{a^2} e^{\frac{-r}{3a}} \sin\theta.\cos\theta.\cos\phi$$

$$= \text{constant} \times \sin\theta.\cos\theta.\sin\phi$$

$$= d_{yz}.$$

$$\psi_{3,2,2} = \frac{1}{81\sqrt{(2\pi a^3)}} \frac{r^2}{a^2} e^{\frac{-r}{3a}} \sin\theta.\cos 2\phi$$

$$= \text{constant} \times \sin^2\theta.\cos 2\theta$$

$$= \text{constant} \times \left(\sin^2\theta.\cos^2\phi - \sin^2\theta.\sin^2\phi\right)$$

$$= d_{x^2-y^2}$$

$$\psi_{3,2,0} = \frac{1}{81\sqrt{(6\pi a^3)}} \frac{r^2}{a^2} e^{\frac{-r}{3a}} \left(\cos^2\theta - 1\right)$$

$$= \text{constant} \times \left(\cos^2\theta - 1\right)$$

$$\psi_{3,2,2} = \frac{1}{81\sqrt{(2\pi a^3)}} \frac{r^2}{a^2} e^{\frac{-r}{3a}} \sin\theta.\cos 2\phi$$

$$\psi_{3,2,2} = \text{Constant} \times \left(\sin^2\theta \sin^2 2\phi\right)$$

$$= \text{Constant} \times \left(\sin\theta\sin\phi.\sin\theta\cos\phi\right)$$

$$= d_{xy}.$$

CHAPTER 6

Approximation Methods

6.1 Perturbation Theory

Perturbation theory expresses the solution to one problem in terms of another problem that has been solved previously.

Suppose we wish to solve the Schrödinger equation,

$$H\psi = E\psi$$

for a particular system, but it is not always possible to get an exact solution similar to those obtained for harmonic oscillator, rigid rotator and hydrogen atom. It turns out that most systems cannot be solved exactly. Two specific examples are the helium atom and the an-harmonic oscillator.

The Hamiltonian operator for the helium atom is,

$$\hat{H} = -\frac{\hbar^2}{2m}\left(\nabla_1^2 + \nabla_2^2\right) - \frac{2e^2}{4\pi\varepsilon_0}\left(\frac{1}{r_1} + \frac{1}{r_2}\right) + \frac{e^2}{4\pi\varepsilon_0}\frac{1}{r_{12}} \qquad (6.1.1)$$

Eq. 6.1.1 can be written in the form,

$$\hat{H} = \hat{H}_{H(1)} + \hat{H}_{H(2)} + \frac{e^2}{4\pi\varepsilon_0}\frac{1}{r_{12}} \qquad (6.1.2)$$

where, $\quad \hat{H}_{H(j)} = -\frac{\hbar^2}{2m}\left(\nabla_j^2\right) - \frac{2e^2}{4\pi\varepsilon_0}\left(\frac{1}{r_j}\right) \quad j = 1 \text{ or } 2 \qquad (6.1.3)$

is the Hamiltonian operator for a single electron around a helium nucleus.

Thus, $\hat{H}_{H(1)}$ and $\hat{H}_{H(2)}$ satisfy the equation,

$$H_{H(j)}\psi_H(r_j,\theta_j,\phi_j) = E_j\psi_H(r_j,\theta_j,\phi_j) \quad j=1 \text{ or } 2 \qquad (6.1.4)$$

where, $\psi_H(r_j,\theta_j,\phi_j)$ is a hydrogen like wave function with $Z = 2$ and where E_j (j = 1 or j = 2) are given by,

$$E_j = -\frac{Z^2 \mu e^4}{8\varepsilon_0^2 h^2 n_j^2} \tag{6.1.5}$$

Notice that if it were not for the inter electronic repulsion term,

$$\frac{e^2}{4\pi\varepsilon_0 r_{12}}$$

in Eq. 6.1.2, the Hamiltonian operator for the helium atom would be separable and the Helium atomic wave functions would be products of hydrogen like atomic wave functions.

Another example of a problem that could be solved readily, if it were not for additional terms in the Hamiltonian, is an-harmonic oscillator. Recall that the harmonic oscillator potential arises naturally as the first term of the Taylor expansion of a general potential about the equilibrium nuclear separation. Consider an an-harmonic oscillator whose potential energy is given by

$$U_{(x)} = \frac{1}{2}k\; x^2 + \frac{1}{6}\gamma\; x^3 + \frac{1}{24}b\; x^4 \tag{6.1.6}$$

The Hamiltonian operator is,

$$\hat{H} = \frac{-\hbar^2}{2\mu}\frac{d^2}{dx^2} + \frac{1}{2}k\; x^2 + \frac{1}{6}\gamma\; x^3 + \frac{1}{24}b\; x^4 \tag{6.1.7}$$

If $\gamma = b = 0$, Eq. 6.1.7 is the Hamiltonian operator for a harmonic oscillator.

The two examples, with their Hamiltonian operators, introduce us to the basic idea behind perturbation theory. In both these cases, the total Hamiltonian consists of two parts, one for which the Schrödinger equation can be solved exactly and an additional term, whose presence prevents an exact solution. We call the first term an unperturbed Hamiltonian and the additional term the perturbation. We shall denote the unperturbed Hamiltonian by $\hat{H}^{(0)}$ and the perturbation by $\hat{H}^{(1)}$ and write,

$$\hat{H} = \hat{H}^{(0)} + \hat{H}^{(1)} \tag{6.1.8}$$

Associated with $\hat{H}^{(0)}$ is a Schrödinger equation, we know how to solve and so we have

$$\hat{H}^{(0)}\psi^{(0)} + E^{(0)}\psi^{(0)} \tag{6.1.9}$$

where $\psi^{(0)}$ and $E^{(0)}$ are the known eigenfunctions and eigenvalues of $\hat{H}^{(0)}$. Eq. 6.1.9 specifies the unperturbed system.

In the case of the helium atom we have,

$$\hat{H}^{(0)} = \hat{H}_{H(1)} + \hat{H}_{H(2)}$$

$$\psi^{(0)} = \psi_H(r_1,\theta_1,\phi_1)\psi_H(r_2,\theta_2,\phi_2) \quad (6.1.10)$$

$$E^0 = \frac{-4\mu e^4}{8\pi\varepsilon_0^2\hbar^2 n_1^2} - \frac{4\mu e^4}{8\pi\varepsilon_0^2\hbar^2 n_2^2}$$

and,

$$\hat{H}^{(1)} = \frac{e^2}{4\pi\varepsilon_0 r_{12}}$$

In the case of an an-harmonic oscillator, we have,

$$\hat{H}^{(0)} = \frac{-\hbar^2}{2\mu}\frac{d^2}{dx^2} + \frac{1}{2}kx^2$$

$$\psi^{(0)} = \frac{(\alpha/\pi)^{1/4}}{(2^n n!)^{1/2}} e^{\frac{-\alpha x^2}{2}} H_n(\alpha^{1/2} x) \quad (6.1.11)$$

$$E^{(0)} = \left(n + \tfrac{1}{2}\right)h\nu \quad \text{and} \quad \hat{H}^{(1)} = \frac{\gamma}{6}x^3 + \frac{b}{24}x^4$$

Intuitively, it can be expected that, if the perturbation terms are not large in some sense, then the solution to the complete perturbed system should be close to the solution to the unperturbed problem. A simple example of this is, when the an-harmonicity terms $\frac{\gamma x^3}{6}$ and $\frac{bx^4}{24}$ are small, we expect the unperturbed system to be perturbed but not altered drastically by the additional term.

Perturbation theory consists of a set of successive corrections to an unperturbed problem

Now we shall derive the equations to the perturbation theory in the lowest level approximation, leaving the higher order terms.

The problem we wish to solve is,

$$\hat{H}\psi = E\psi \quad (6.1.12)$$

where

$$\hat{H} = \hat{H}^{(0)} + \hat{H}^{(1)} \quad (6.1.13)$$

and where the problem

$$\hat{H}^{(0)}\psi^{(0)} = E^{(0)}\psi^{(0)} \tag{6.1.14}$$

has been solved previously, so that $\psi^{(0)}$ and $E^{(0)}$ are known.

Assuming now that the effect of $\hat{H}^{(1)}$ is small, we write,

$$\psi = \psi^{(0)} + \Delta\psi$$

and

$$E = E^{(0)} + \Delta E \tag{6.1.15}$$

where we assume that $\Delta\psi$ and ΔE are small. Substituting these equations in Eq. 6.1.12 we get,

$$H^{(0)}\psi^{(0)} + H^{(1)}\psi^{(0)} + H^{(0)}\Delta\psi + H^{(1)}\Delta\psi = \\ E^{(0)}\psi^{(0)} + \Delta E\psi^{(0)} + E^{(0)}\Delta\psi + \Delta E\Delta\psi \tag{6.1.16}$$

The first term on each side of Eq. 6.1.16 cancel because of Eq. 6.1.14. In addition, we shall neglect the last terms on each side, because they represent the product of two small terms. Eq, 6.1.16 becomes

$$H^{(0)}\Delta\psi + \hat{H}^{(1)}\psi^{(0)} = E^{(0)}\Delta\psi + \Delta E\psi^{(0)} \tag{6.1.17}$$

Realize that $\Delta\psi$ and ΔE are the unknown quantities in this equation.

Note that all the terms in Eq. 6.1.17 are of the same order, in the sense that each is the product of an unperturbed term and a small term. Therefore, we say that the equation is first order in perturbation and we are doing here is the first order perturbation theory. The above neglected second order terms lead to second order corrections. Eq. 6.1.17 can be simplified considerably by multiplying both sides from the left by $\psi^{(0)*}$ and integrating over all space. By doing this and rearranging slightly, we find,

$$\int \psi^{(0)*} \left[\hat{H}^{(0)} - E^{(0)}\right]\Delta\psi \, d\tau + \int \psi^{(0)*} \hat{H}^{(1)} \psi^{(0)} d\tau = \\ \Delta E \int \psi^{(0)*} \psi^{(0)} \, d\tau \tag{6.1.18}$$

The integral in the last term in Eq. 6.1.18 is unity because $\psi^{(0)}$ is taken to be normalized. More importantly, the first term on the left hand side of Eq. 6.1.18 is zero. To see this, remember that $\hat{H}^{(0)} - E^{(0)}$ is Hermitian, and so we have that,

$$\int \psi^{(0)*}\left[\hat{H}^{(0)} - E^{(0)}\right]\Delta\psi d\tau = \int \left\{\left[\hat{H}^{(0)} - E^{(0)}\right]\psi^{(0)}\right\}^* \Delta\psi d\tau \tag{6.1.19}$$

But according to Eq. 6.1.14, the integrand here vanishes. Thus, Eq. 6.1.18 becomes,

$$\Delta E = \int \psi^{(0)*} \hat{H}^{(1)} \psi^{(0)} d\tau \tag{6.1.20}$$

This equation is called the first order correction to energy $E_{(0)}$.

To the first order, the energy is,

$$E = E^{(0)} + \int \psi^{(0)*} H^{(1)} \psi^{(0)} d\tau + \text{Higher order terms} \tag{6.1.21}$$

6.2 The Variational Method

The variational method provides an upper bound to the ground state energy of a system.

The second approximation method that we shall discuss is more useful than perturbation theory, because it does not require that there be a similar problem that has been solved previously. This second approximation method is the variational method.

Consider the ground state of some particular arbitrary system. The ground state wave function, ψ_0 and energy $E_{(0)}$ satisfy the Schrödinger equation.

$$\hat{H}\psi_{(0)} = E_{(0)}\psi_{(0)} \tag{6.2.1}$$

Multiply Eq. 6.2.1 from the left by $\psi^*_{(0)}$ and integrate overall space to obtain

$$E_{(0)} = \frac{\int \psi^*_{(0)} \hat{H} \psi_{(0)} d\tau}{\int \psi^*_{(0)} \psi_{(0)} d\tau} \tag{6.2.2}$$

where, $d\tau$ represents approximate volume element. We have not set the denominator to unity in Eq. 6.2.2, in order to allow the possibility that $\psi^{(0)}$ is not normalized beforehand. There is a beautiful theorem that says that if we substitute any other function for $\psi^{(0)}$ in Eq. 6.2. 2 and calculate,

$$E_\phi = \frac{\int \phi^* \hat{H} \phi d\tau}{\int \phi^* \phi d\tau} \tag{6.2.3}$$

the E_ϕ calculated through Eq. 6.2.3 will be greater than the ground state energy, $E_{(0)}$. In an equation, we have the variational principle

$$E_\phi \geq E_0 \tag{6.2.4}$$

The variational principle says that we can calculate an upper bound on E_0 by using any trial function we wish. The closure ϕ is to $\psi_{(0)}$ in some sense; the closure E_ϕ will be to E_0. We can choose such that it depends on some arbitrary parameters α, β, γ, called variational parameters. The energy also will depend on these variational parameters and Eq. 6.2.4 will read,

$$E_\phi(\alpha,\beta,\gamma.....) \geq E_0 \quad (6.2.5)$$

Now we can minimize E_ϕ with respect to each of the variational parameters and thus approach the exact ground state energy E_0.

As a specific example, consider the ground state of the hydrogen atom. Although we can solve this problem exactly, let us assume that we cannot and use the variational method. We shall compare the variational result with the exact result.

Because $l = 0$ in the ground state, the Hamiltonian operator is,

$$\hat{H} = \frac{-\hbar^2}{2\mu r^2}\frac{d}{dr}\left(r^2\frac{d}{dr}\right) - \frac{e^2}{4\pi\varepsilon_0 r} \quad (6.2.6)$$

As a trial function, we shall try a Gaussian function of the form $\phi(r) = e^{-\alpha r^2}$ where, α is a variational parameter. It can be shown that,

$$4\pi\int_0^\infty dr\, r^2 \phi^*(r)\hat{H}\phi(r) = \frac{3\hbar^2\pi^{3/2}}{\left(4\sqrt{2}\mu\alpha^{1/2}\right)} - \frac{e^2}{4\varepsilon_0\alpha}$$

and that

$$4\pi\int_0^\infty dr\, r^2 \phi^*(r)\phi(r) = \left(\frac{\pi}{2\alpha}\right)^{3/2}$$

Therefore, from Eq. 6.2.3,

$$E(\alpha) = \frac{3\hbar^2\alpha}{2\mu} - \frac{e^2\alpha^{1/2}}{2^{1/2}\varepsilon_0\pi^{3/2}} \quad (6.2.7)$$

Now differentiate $E(\alpha)$ with respect to α and set the result equal to zero to find,

$$\alpha = \frac{\mu^2 e^4}{18\pi^3\varepsilon_0^2\hbar^4} \quad (6.2.8)$$

as the value of 'α' that minimizes $E(\alpha)$. Substituting Eq. 6.2.8 back into Eq. 6.2.7 we find that,

$$E_{min} = -\frac{4}{3\pi}\left(\frac{\mu e^4}{16\pi^2\varepsilon_0^2\hbar^2}\right) = -0.424\left(\frac{\mu e^4}{16\pi^2\varepsilon_0^2\hbar^2}\right) \quad (6.2.9)$$

compared to the exact value,

$$E_0 = -\frac{1}{2}\left(\frac{\mu e^4}{16\pi^2\varepsilon_0^2\hbar^2}\right) = -0.5\left(\frac{\mu e^4}{16\pi^2\varepsilon_0^2\hbar^2}\right) \quad (6.2.10)$$

Note that $E_{min} > E_0$ and the variational theorem assures us. Thus, we see that the variational method gives a rather good result. We can obtain a better result by using a more flexible trial function.

Proof

Now that we have illustrated the utility of the variational theorem by example, we shall prove it. Let,

$$\hat{H}\psi_n = E_n\psi_n \quad (6.2.11)$$

be the problem of interest. Even though we do not know explicitly, we do know that we can expand any suitable arbitrary function ϕ in terms of the ψ_n and write,

$$\phi = \sum_n c_n\psi_n \quad (6.2.12)$$

If we substitute this into Eq. 6.2.3 and use the fact that ψ_n are orthonormal, then we obtain,

$$E_\phi = \frac{\sum_n c_n^* c_n E_n}{\sum_n c_n^* c_n} \quad (6.2.13)$$

Subtract E_0 from the left-hand side and from the right hand side to find,

$$E_\phi - E_0 = \frac{\sum_n c_n^* c_n (E_n - E_0)}{\sum_n c_n^* c_n} \quad (6.2.14)$$

Now by definition, E_0 is the ground-state energy. Consequently, $E_n - E_0 \geq 0$ for all values of n, and because all the $c_n c_n^* > 0$, Eq-6.2.14 shows that,

$$E_\phi - E_0 \geq 0 \quad (6.2.15)$$

which is the variational theorem.

6.3 The Hartree Theory

Compared to the simplified one-dimensional systems or even to the one electron atom, multi-electron atoms are quite complicated. But, it is possible to treat them in a reasonable way using a succession of approximations. Only the most important interactions experienced by the atomic electrons are treated in the first approximation, and then the treatment is made more exact in succeeding approximations that take into account the less important interactions. In this way, the treatment is broken into a series of steps, none of which is too difficult. The results obtained will certainly justify the effort expended, because we shall have a detailed understanding of the atoms that are the constituents of everything in this universe.

In the first approximation used in treating a multi-electron atom of atomic number z, we must consider the coulomb interaction between each of its 'z' electrons of charge 'e' and its nucleus of charge 'ze'. Due to the magnitude of the nuclear charge, this is the strongest single interaction felt by each electron. But, even in the first approximation we must also consider the coulomb interactions between each electron and all the other electrons in the atom. These interactions are individually weaker than the interaction between each electron and the nucleus, but they are certainly not negligible. Furthermore, in a typical multi-electron atom, there are so many interactions between an electron and all other electrons that their net effect is very strong, except if the electron is quite near the nucleus.

Surface of the Atom

 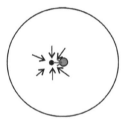

Fig. 6.3.1(Left) The strong attractive force (big arrow) exerted by the nucleus on an electron near the surface of the atom and the weak repulsive forces (small arrows) exerted by the other electrons. The net effective repulsive force is important which is the reinforced one of all the forces.

Fig. 6.3.1(Right) The very strong attractive force (arrow pointing towards the nucleus) exerted by the nucleus on an electron near the centre of the atom and the weak repulsive forces (arrows pointing towards the electron) exerted by the other electrons. Here the repulsive forces tend to cancel each other.

On the other hand, the first approximation must not be so complicated that the Schrödinger equation to which it leads is unsolvable. In practice, this requirement means that in the first approximation the atomic electrons must be treated as moving independently, so that the motion of one electron does not depend on the motion of the others. Then the time independent Schrödinger equation for the system can be separated into a set of equations, one for each electron, which can be solved without too much difficulty, since each involves the coordinates of a single electron only.

The requirement of the last two paragraphs are in conflict - the coulomb interactions between the electrons must be considered, but the electrons must be treated as moving independently. A compromise between the requirements is obtained by assuming each electron to move independently in a spherically symmetrical net potential V(r), where 'r' is the radial coordinate of the electron with respect to the nucleus. The net potential is the sum of the spherically symmetrical attractive coulomb potential due to the nucleus and a spherically symmetrical repulsive potential, which represents the average effect of the repulsive coulomb interactions between a typical electron and its Z-1 colleagues. It can be seen from the figure that very near the centre of the atom the behavior of the net potential acting on an electron should be essentially like that of the coulomb potential due to the nuclear charge + ze. The reason is that in this region, the interactions of the electron with the other electrons tend to cancel. It can also be seen from the figure that very far from the centre, the behavior of the net potential should be essentially like that of the coulomb potential due to net charge + e, which represents the nuclear charge + ze shielded by the charge -(Z-1)e of other electrons.

It might be seen that there is no way to find the net potential of an atom at intermediate distances from its centre. The problem is that it obviously depends on the details of the charge distribution of the atomic electrons, and this is not known until solutions have been obtained to the Schrödinger equation that contains the net potential. However, it can be taken care of by demanding that the net potential be self-consistent. That is, we calculate the electron charge distribution from the correct net potential, and then evaluate the net potential from the charge distribution. We demand that the potential with which we end up must be the same as the potential with which we started. As we shall see, this condition of self-consistency is enough to determine the correct net potential.

Most of the work in this field, has been started by Douglas Hartree and Collaborators in 1928. It involves solving the time-independent Schrödinger equation for a system of 'z' electrons moving independently in the atom. The total potential of the atom can be written as a sum of a set of Z identical net potentials V(r), each depending on the radial coordinate 'r' of one electron only. Consequently, the equation can be separated into a set of z time independent Schrödinger equations, all of which are of the same form, and each of which describes one electron moving independently in its net potential. A typical time independent Schrödinger equation for one electron is,

$$\frac{\hbar^2}{2m}\nabla^2 \psi_{(r,\theta,\phi)} + V\psi_{(r,\theta,\phi)} = E\psi_{(r,\theta,\phi)}$$

Here, r, θ, ϕ are the spherical polar coordinates of the typical electron, ∇^2 is the Laplacian operator in these coordinates; E is the total energy of the electron; V(r) is its net potential; and $\psi_{(r,\theta,\phi)}$ is the eigenfunction of the electron. The total energy of the atom is the sum of Z of these total energies. The total eigenfunction for the atom is composed of the product of Z of these eigenfunctions that describe the independently moving electrons.

Initially the exact form of the net potential V(r) experienced by the typical electron is not known, but it can be found by going through a self-consistent treatment comprised of the following steps.

1. A first guess at the form of V(r) is obtained by taking,

$$V_{(r)} = \frac{-Ze^2}{4\pi\varepsilon_0 r} \qquad r \to 0$$

$$V_{(r)} = \frac{-e^2}{4\pi\varepsilon_0 r} \qquad r \to \infty$$

and by taking any reasonable interpolation of intermediate values of 'r'. This guess is based on the idea, mentioned previously, that an electron very near the nucleus feels the full coulomb attraction of its charge +ze, while an electron very far from the nucleus feels a net charge of +e because the nuclear charge is shielded by the charge -(Z-1)e of the other electrons surrounding the nucleus.

2. The time independent Schrödinger equation for a typical electron is solved for the net potential V(r) obtained in the previous step. This is not easy to do because the radial part of the equation must be solved by numerical integration, since V(r) is a complicated function. The eigenfunctions for a typical electron found in this step are: $\psi_{\alpha(r,\theta,\phi)}, \psi_{\beta(r,\theta,\phi)}, \psi_{\nu(r,\theta,\phi)}, \ldots$ They are listed in order of increasing energy of the corresponding eigenvalues: $E_\alpha, E_\beta, E_\nu, \ldots$ Each of these symbols $\alpha, \beta, \nu, \ldots$ stands for a complete set of three space and one spin quantum numbers for the electron.

3. To obtain the ground state of the atom, the quantum states of its electrons are filled in such a way as to minimize the total energy and yet satisfy the weaker condition of the exclusion principle. That is, the states are filled in order of increasing energy, with one electron in each state as illustrated in

Fig. 6.3.2. Then the eigenfunction for the first electron will be $\psi_\alpha(r_1,\theta_1,\phi_1)$, the eigenfunction for the second will be $\psi_\alpha(r_2,\theta_2,\phi_2)$, and so forth through the z eigenfunctions corresponding to the z lowest eigenvalues, obtained in the previous step.

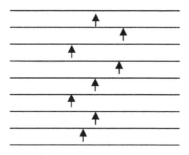

Fig: 6.3.2 Schematic energy level diagram illustrating the effect of the exclusion principle in limiting the population of each quantum state of an atom.

4. The electron charge distributions of the atom are then evaluated from the eigenfunctions specified in the previous step. This is done by taking the charge distribution for each electron as the product of its charge 'e' times its probability density function $\psi^*\psi$. The justification is that $\psi^*\psi$ determines the probability that the charge distributions of 'Z-1' representative electrons are added to the nuclear charge distribution, a point charge +ze at the origin to determine the total charge distribution of the atom as seen by a typical electron.

5. Gauss's law of electrostatics is used to calculate the electric field produced by the total charge distribution obtained in the previous step. The integral of this electric field is then evaluated to obtain a more accurate estimate of the net potential, V(r) experienced by a typical electron. The V(r) that is found, generally differs from the estimate made in step 1.

6. If it is apparently different, the entire procedure is repeated, starting at step 2 and using the new V(r). After several cycles (2 —> 3 —> 4 —> 5 —> 2 —> 3 —> 4 —> —5), the V(r) obtained at the end of a cycle is essentially the same as that used in the beginning. Then this V(r) is the self-consistent net potential, and the eigenfunctions calculated from this potential describe the electrons in the ground state of the multi-electron atom.

In the Hartree procedure, the weaker condition of the exclusion principle is satisfied by the requirement of the step 3 that only one electron populates each quantum state. But, the stronger condition is not satisfied since antisymmetric total eigenfunctions are not used. The reason is that an antisymmetric eigenfunction would involve a linear combination of Z! =

Z(Z-1)(Z-2) ....1 terms, which is an extremely large number for all atoms except for those of very small Z. The procedure is difficult and the use of antisymmetric eigenfunctions would make it even more difficult. Anyway, the main effect of using antisymmetric total eigenfunctions would be to decrease the separation between certain parts of electrons, and increase it between others. This leaves the average electron charge distribution of the atom essentially unchanged. Since the average electron charge distribution is the important quantity in the approximation treated by Hartree, the use of eigenfunctions which are not of a definite symmetry does not introduce a significant error. This has been verified by Fock. He made calculations using antisymmetric total eigenfunctions for a restricted selection of atoms, and he compared his results with those obtained by Hartree. When the excited states of the atoms are discussed, it will be necessary to take into account the fact that antisymmetric total eigenfunctions must be used to give a completely accurate description of a system of electrons. Fock's calculations are feasible because, it is only necessary to anti-symmetrize the part of the total eigenfunction describing the behavior of a limited number of electrons in a "partially filled sub-shell".

6.4 Exercises

Q1: Calculate the ground state energy of a harmonic oscillator (One dimensional)

Let the trial function $= \psi e^{-cx^2}$ be where 'c' is an arbitrary constant.

$$E = \frac{\int \psi^* \hat{H} \psi dx}{\int \psi^* \psi dx}$$

But for harmonic oscillator is

$$-\frac{\hbar^2}{2m}\frac{d^2}{dx^2} + \frac{1}{2}kx^2$$

Before doing the evaluation of the integrals first normalize the trial wave function.

(A = Normalization constant)

$$\therefore \int_{-\infty}^{+\infty} \psi^* \psi dx = 2A^2 \int_0^{\infty} e^{-2cx^2} dx = 1$$

$$= 2A^2 \left(\frac{1}{2}\sqrt{\frac{\pi}{2c}} \right)$$

$$\therefore A^2 = \pm \left(\frac{2c}{\pi} \right)^{\frac{1}{2}}$$

$$\therefore \int_{-\infty}^{+\infty} A\psi^* A\psi dx = 1$$

$$\therefore E = \int_{-\infty}^{+\infty} Ae^{-cx^2}\left[-\frac{\hbar^2}{2m}\frac{d^2}{dx^2}+\frac{1}{2}kx^2\right]Ae^{-cx^2}dx$$

$$= 2\int_0^{\infty} Ae^{-cx^2}\left[-\frac{\hbar^2}{2m}\frac{d^2}{dx^2}+\frac{1}{2}kx^2\right]Ae^{-cx^2}dx$$

$$= -\frac{\hbar^2 A^2}{m}\int_0^{\infty} e^{-cx^2}\frac{d^2}{dx^2}e^{-cx^2}dx + A^2k\int_0^{\infty} x^2 e^{-2cx^2}dx$$

$$= -\frac{\hbar^2 A^2}{m}\int_0^{\infty} e^{-cx^2}\left(-2ce^{-cx^2}+4c^2x^2 e^{-cx^2}\right)dx + A^2k\int_0^{\infty} x^2 e^{-2cx^2}dx$$

$$= -\frac{\hbar^2 A^2}{m}\left[\int_0^{\infty}-2ce^{-2cx^2}dx + \int_0^{\infty}4c^2x^2 e^{-2cx^2}dx\right] + A^2k\int_0^{\infty} x^2 e^{-2cx^2}dx$$

$$= -\frac{\hbar^2 A^2}{m}\left[(2c)\frac{1}{2}\sqrt{\frac{\pi}{2c}}+4c^2\frac{1}{4}\sqrt{\frac{\pi}{8c^3}}\right] + A^2k\left(\frac{1}{4}\sqrt{\frac{\pi}{8c^3}}\right)$$

but

$$A = \sqrt{\frac{2c}{\pi}}$$

$$\therefore E = -\frac{\hbar^2}{m}\left[(-c)\sqrt{\frac{\pi}{2c}}\sqrt{\frac{2c}{\pi}}+c^2\sqrt{\frac{\pi}{8c^3}}\sqrt{\frac{2c}{\pi}}\right]+k\left(\frac{1}{4}\sqrt{\frac{\pi}{8c^3}}\sqrt{\frac{2c}{\pi}}\right)$$

$$= \frac{\hbar^2}{m}\left(c-\frac{c}{2}\right)+\frac{k}{8c} = \frac{\hbar^2}{2m}c+\frac{k}{8c}$$

$$\frac{\partial E_T}{\partial c} = \frac{\hbar^2}{2m}-\frac{k}{8c^2}=0$$

$$\therefore \frac{\hbar^2}{2m} = \frac{k}{8c^2}$$

$$c = \frac{\pi}{h}\sqrt{mk}$$

But

$$A = \sqrt{\frac{2c}{\pi}}$$

$$\therefore \frac{\hbar^2}{2m} = \frac{k}{8c^2}$$

$$c = \frac{\Pi}{h}\sqrt{mk}$$

$$E_{min} = \frac{\hbar^2}{2m}c + \frac{k}{8c} = \frac{h^2 c}{8\pi^2 m} + \frac{k}{8c} = \frac{h^2 c^2 + \pi^2 mk}{8\pi^2 mc}$$

$$= \frac{\sqrt{mk}(2\pi^2 h)}{8\pi^3 m} = \frac{h}{4\pi}\sqrt{\frac{k}{m}}$$

but

$$\frac{1}{2\pi}\sqrt{\frac{k}{m}} = \nu_0$$

$$\therefore E_{min} = \tfrac{1}{2} h\nu_0$$

(Ground state energy of the harmonic oscillator which is also the zero point energy)

Special integrals

$$\int_0^\infty e^{-kx^2} dx = \tfrac{1}{2}\sqrt{\frac{\pi}{k}}$$

$$\int_0^\infty xe^{-kx^2} dx = \frac{1}{2k}$$

$$\int_0^\infty x^2 e^{-kx^2} dx = \tfrac{1}{4}\sqrt{\frac{\pi}{k^3}}$$

*Calculation

$$\frac{d^2}{dx^2}e^{-cx^2} = \frac{d}{dx}\left[-2cxe^{-cx^2}\right]$$

$$= -2ce^{-cx^2} + 4c^2 x^2 e^{-cx^2}$$

CHAPTER 7

Bonding in Molecules

7.1 Molecular Orbital Theory

LCAO Method – H_2^+ - ion

Atom is a centrosymmetric system. However, in molecules it will be disturbed. We shall make a reasonable approximation that a molecular orbital (MO) is a Linear Combination of Atomic Orbitals (LCAO).

As individual hydrogen atoms at quite large distance from each other, (theoretically) are brought closer and closer, the nucleus of each atom will start to attract the electrons originally associated with the other atom. The change in energy of the system, as a function of distance, is usually shown in the form of a curve called a Morse curve.

When the distance separating the nuclei is at or near the bonding distance, the electron in the system is associated with the two nuclei instead of the original atomic orbitals on each atom. The electron is now associated with a molecular orbital(MO) that is the combination of the two atomic orbitals.

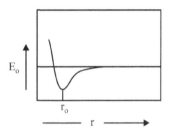

Fig. 7.1.1 Change of energy as a function of distance.

When the electron is near one nucleus, the MO may be assumed to resemble the atomic orbital of the atom. Let the wave function be ϕ_1. Similarly when the electron is in the neighborhood of the other nucleus the MO resembles the atomic orbital of the other atom. The wave function for this is given by ϕ_2. Since the complete MO has the characteristics separately possessed by ϕ_1 and ϕ_2, the total wave function for the MO is formed by the linear combination of atomic orbitals. Linear combinations are made by simple addition or subtraction of the functions to be combined. In this case it is

$$\psi_{mol} = \phi_1 + \phi_2$$

This terminology, called the Liner Combination of Atomic Orbitals or LCAO method, was first suggested by R. S. Mulliken.

The Schrödinger equation can be written as

$$\nabla^2 \psi + \frac{8\pi^2 m}{h^2}(E-V)\psi = 0$$

This can be rearranged as follows:

$$-\frac{h^2}{8\pi^2 m}\nabla^2\psi + V\psi = E\psi$$

or

$$\left[V - \frac{h^2}{8\pi^2 m}\left(\frac{\partial^2}{\partial x^2} + \frac{\partial^2}{\partial y^2} + \frac{\partial^2}{\partial z^2}\right)\right]\psi = E\psi \quad (7.1.1)$$

The left side of the equation can be considered as the action of an operator called the Hamiltonian operator on ψ, where ψ is now a molecular orbital. We can abbreviate the expression as

$$H\psi = E\psi \quad (7.1.2)$$

where H is the Hamiltonian operator. Just as in the case of atomic wave functions, the MO wave function ψ can be either positive or negative and ψ^2 is a quantity proportional to electron density. If we multiply both sides of equation Eq. 7.1.2 by ψ and integrate over all space, we obtain

$$\int \psi H \psi d\tau = E \int \psi^2 d\tau \quad (7.1.3)$$

In Eq.7.1.3, we have allowed a single integral sign to stand for a triple integral sign and have made substitution $d\tau = d_x d_y d_z$. By rearrangement, we obtain the following expression for energy.

$$E = \frac{\int \psi H \psi d\tau}{\int \psi^2 d\tau} \quad (7.1.4)$$

The wave function is represented by the following linear function

$$\psi_{mol} = c_1 \phi_1 + c_2 \phi_2 \quad (7.1.5)$$

where ϕ_1 and ϕ_2 are the atomic orbital wave functions of atoms 1 and 2, and c_1 and c_2 are coefficients to be determined. Substituting the value of the wave function in Eq. 7.1.4.

Chapter 7 | Bonding in Molecules   117

$$E = \frac{\int (c_1\phi_1 + c_2\phi_2) H (c_1\phi_1 + c_2\phi_2) d\tau}{\int (c_1\phi_1 + c_2\phi_2)(c_1\phi_1 + c_2\phi_2) d\tau}$$

$$= \frac{\int c_1\phi_1 H c_1\phi_1 d\tau + \int c_1\phi_1 H c_2\phi_2 d\tau + \int c_2\phi_2 H c_1\phi_1 d\tau + \int c_2\phi_2 H c_2\phi_2 d\tau}{\int c_1^2 \phi_1^2 d\tau + 2\int c_1 c_2 \phi_1 \phi_2 d\tau + \int c_2^2 \phi_2^2 d\tau}$$

$$= \frac{c_1^2 \int \phi_1 H \phi_1 d\tau + c_1 c_2 \int \phi_1 H \phi_2 d\tau + c_1 c_2 \int \phi_2 H \phi_1 d\tau + c_2^2 \int \phi_2 H \phi_2 d\tau}{c_1^2 \int \phi_1^2 d\tau + 2 c_1 c_2 \int \phi_1 \phi_2 d\tau + c_2^2 \int \phi_2^2 d\tau}$$

But we know that $Hc_1\phi_1 = c_1 H\phi_1$

and $\int \phi_1 H \phi_2 d\tau = \int \phi_2 H \phi_1 d\tau$

Therefore we may write

$$E = \frac{c_1^2 \int \phi_1 H \phi_1 d\tau + 2 c_1 c_2 \int \phi_1 H \phi_2 d\tau + c_2^2 \int \phi_2 H \phi_2 d\tau}{c_1^2 \int \phi_1^2 d\tau + 2 c_1 c_2 \int \phi_1 \phi_2 d\tau + c_2^2 \int \phi_2^2 d\tau}$$

For simplification, we make the following substitutions

$$H_{11} = \int \phi_1 H \phi_1 d\tau$$

$$H_{22} = \int \phi_2 H \phi_2 d\tau$$

$$H_{12} = \int \phi_1 H \phi_2 d\tau$$

$$S_{11} = \int \phi_1^2 d\tau$$

$$S_{22} = \int \phi_2^2 d\tau$$

$$S_{11} = \int \phi_1 \phi_2 d\tau$$

Hence,

$$E = \frac{c_1^2 H_{11} + 2 c_1 c_2 H_{12} + c_2^2 H_{22}}{c_1^2 S_{11} + 2 c_1 c_2 S_{12} + c_2^2 S_{22}} \quad (7.1.6)$$

Since we desire the minimum value of E, it is necessary to minimize E with respect to both c_1 and c_2. Therefore, it is necessary that, $\dfrac{\partial E}{\partial c_1} = \dfrac{\partial E}{\partial c_2} = 0$.

Differentiating Eq. 7.1.6 with respect to c_1

$$\frac{\partial E}{\partial C_1} = \frac{(c_1^2 S_{11} + 2 c_1 c_2 S_{12} + c_2^2 S_{22})(2 c_1 H_{11} + 2 c_2 H_{12}) - (c_1^2 H_{11} + 2 c_1 c_2 H_{12} + c_2^2 H_{22})(2 c_1 S_{11} + 2 c_2 S_{12})}{(c_1^2 S_{11} + 2 c_1 c_2 S_{12} + c_2^2 S_{22})^2} = 0$$

This is rearranged as

$$\frac{\left(c_1^2 S_{11} + 2c_1c_2 S_{12} + c_2^2 S_{22}\right)\left(2c_1 H_{11} + 2c_2 H_{12}\right)}{\left(c_1^2 S_{11} + 2c_1c_2 S_{12} + c_2^2 S_{22}\right)^2}$$

$$= \frac{\left(c_1^2 H_{11} + 2c_1c_2 H_{12} + c_2^2 H_{22}\right)\left(2c_1 S_{11} + 2c_2 S_{12}\right)}{\left(c_1^2 S_{11} + 2c_1c_2 S_{12} + c_2^2 S_{22}\right)^2}$$

But $$E = \frac{\left(c_1^2 H_{11} + 2c_1c_2 H_{12} + c_2^2 H_{22}\right)}{\left(c_1^2 S_{11} + 2c_1c_2 S_{12} + c_2^2 S_{22}\right)}$$

So the above equation becomes

$$\frac{\left(2c_1 H_{11} + 2c_2 H_{12}\right)}{c_1^2 S_{11} + 2c_1c_2 S_{12} + c_2^2 S_{22}} = \frac{E\left(2c_1 S_{11} + 2c_2 S_{12}\right)}{c_1^2 S_{11} + 2c_1c_2 S_{12} + c_2^2 S_{22}}$$

that is

$$c_1 H_{11} + c_2 H_{12} = E\left(c_1 S_{11} + c_2 S_{12}\right)$$

or

$$c_1\left(H_{11} - ES_{11}\right) + c_2\left(H_{12} - ES_{12}\right) = 0 \qquad (7.1.7)$$

Similarly differentiating Eq. 7.1.6 with respect to c_2 we obtain the following equation

$$c_1\left(H_{12} - ES_{12}\right) + c_2\left(H_{22} - ES_{22}\right) = 0 \qquad (7.1.8)$$

Eqs. 7.1.7 and 7.1.8 are called the secular equations.

It is to be noted that the secular equations are of the form

$$ax + by = 0$$
$$cx + dy = 0$$

and if we solve this set of linear homogeneous equations, we see that

$$(ad - bc)y = 0$$

In order this equation to be valid, it is apparent that either 'y' is zero or else the coefficient of 'y' is zero. If 'y' is zero, no problem really exists. Therefore, a nontrivial solution requires that the coefficient of 'y' be zero. This can be expressed as $(ad - bc) = 0$.

The same condition applied to the secular Eqs. 7.1.7 and 7.1.8, where c_1 and c_2 are not equal to zero. If they are zero then, $\psi = c_1\phi_1 + c_2\phi_2 = 0$ which is meaningless. Hence, the coefficients of c_1 and c_2 must be zero in order to have a nontrivial solution.

Chapter 7 | Bonding in Molecules

The secular equations are

$$c_1(H_{11} - ES_{11}) + c_2(H_{12} - ES_{12}) = 0$$
$$c_1(H_{12} - ES_{12}) + c_2(H_{22} - ES_{22}) = 0$$

A nontrivial solution to these equations can be expressed in terms of the secular determinant.

$$(H_{11} - ES_{11})(H_{22} - ES_{22}) - (H_{12} - ES_{12})^2 = 0 \qquad (7.1.9)$$

The terms H_{11} and H_{22} are called "coulomb integrals". Coulomb integral is apparently the energy of an electron in the valence atomic orbital, α.

At least this approximation is reasonable for a neutral molecule, in which electron - electron and nucleus – nucleus repulsions somewhat compensate. Hence, we may write $H_{11} = \alpha_1$ and $H_{22} = \alpha_2$. The term H_{12} is called the resonance integral, β and is essentially the interaction energy of the two atomic orbitals which is also called the covalent integral. Both α and β have negative values.

If the atomic orbital wave functions in Eq.7.1.5 are normalized, then

$$S_{11} = \int \phi_1^2 d\tau = S_{22} = \int \phi_2^2 d\tau \qquad (7.1.10)$$

Eq.7.1.10 simply states that the probability of finding an electron in the orbital is exactly unity. The term S_{12} is called the "overlap integral" because it is a measure of the extent to which orbitals 1 and 2 overlap. For simplification we shall omit the subscripts and write S for the overlap integral. Then the secular determinant will be reduced to

$$\begin{vmatrix} \alpha_1 - E & \beta - ES \\ \beta - ES & \alpha_2 - E \end{vmatrix} = 0$$

For a homo-nuclear species such as H_2^+, we may substitute $\alpha_1 = \alpha_2 = \alpha$. The detrimental equation the corresponds to

$$\begin{vmatrix} \alpha - E & \beta - ES \\ \beta - ES & \alpha - E \end{vmatrix} = 0$$

that is
$$(\alpha - E)^2 - (\beta - ES)^2 = 0 \qquad (7.1.11)$$

Solution:

Eq.7.1.11 can be written as

$$(\alpha - E)^2 - (\beta - ES)^2 = 0$$

$$(\alpha - E) = \pm(\beta - ES) \qquad (7.1.12)$$

If $(\alpha - E) = -(\beta - ES)$

then $E = \dfrac{\alpha + \beta}{1 + S}$ and $\qquad (7.1.13)$

if $(\alpha - E) = +(\beta - ES)$

then $E = \dfrac{\alpha - \beta}{1 - S} \qquad (7.1.14)$

Eqs. 7.1.13 and 7.1.14 denote symmetric and antisymmetric energy states.
By taking the secular Eq.7.1.7

$$c_1(H_{11} - ES_{11}) + c_2(H_2 - ES_{12}) = 0$$

and by appropriate substitution, we obtain

$$c_1(\alpha - E) = -c_2(\beta - ES)$$

Therefore,

$$c_1 = -\dfrac{\beta - ES}{\alpha - E}c_2$$

From this relation it can be seen that, when

$E = \dfrac{\alpha + \beta}{1 + S}$, then $c_1 = c_2$ and when

$E = \dfrac{\alpha - \beta}{1 - S}$, then $c_1 = -c_2$.

Thus, the molecular orbital wave function can be written as

$$\psi = c_1\phi_1 \pm c_1\phi_2.$$

To evaluate c_1 we must normalize the wave function.

$$\int \psi^2 d\tau = \int (c_1\phi_1 \pm c_1\phi_2)^2 d\tau = 1$$

$$c_1^2 \int \phi_1^2 d\tau \pm 2c_1^2 \int \phi_1\phi_2 d\tau + c_1^2 \int \phi_2^2 d\tau = 1$$

$$c_1^2 S_{11} \pm 2c_1^2 S + c_1^2 S_{22} = 1 \text{ but } S_{11} = S_{22} = 1.$$

Hence, the above equation becomes

$$c_1^2 \pm 2c_1^2 S + c_1^2 = 1$$

$$2c_1^2 \pm 2c_1^2 S = 1$$

$$\therefore \quad c_1 = \pm\dfrac{1}{\sqrt{2 \pm 2S}} \qquad (7.1.15)$$

Arbitrarily taking the positive sign of the normalization constant, the positive sign under the radical sign corresponds to $c_1 = c_2$ and the negative sign under the radical sign corresponds to $c_1 = -c_2$. Obviously, the following wave functions are normalized.

$$\psi_B = \frac{1}{\sqrt{2 \pm 2S}}(\phi_1 + \phi_2) \quad (7.1.16)$$

$$\psi_A = \frac{1}{\sqrt{2 - 2S}}(\phi_1 - \phi_2) \quad (7.1.17)$$

The valence electron density is obtained by squaring these functions

$$\psi_B^2 = \frac{1}{2+2S}(\phi_1^2 + \phi_2^2 + 2\phi_1\phi_2)$$

$$\psi_A^2 = \frac{1}{2-2S}(\phi_1^2 + \phi_2^2 - 2\phi_1\phi_2)$$

ψ_B^2 shows an increase in electron density in the region of overlap between the atoms over that of the individual atoms. Such an electron distribution stabilizes the system, and we refer to ψ_B as the "bonding" MO. The energy level of is given by $E = \frac{\alpha + \beta}{1 + S}$. ψ_A^2 shows a decrease in electron density in the overlap region, and the system is unstable relative to the separate atoms.

We refer to ψ_A as the "anti-bonding" MO for which $E = \frac{\alpha - \beta}{1 - S}$.

In figure it show a plot of ϕ_1^2, ϕ_2^2, ψ_b^2 and ψ_a^2 along the inter nuclear line.

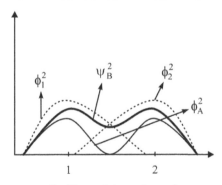

Position on inter nuclear axis

Fig.7.1.2 Plot of electron densities for the orbitals ϕ_1, ϕ_2, ψ_B and ψ_A along the internuclear axis of H_2^+.

The dashed lines indicate ϕ_1^2 and ϕ_2^2, that is the electron density of the individual atomic orbitals. The lower solid line indicates ψ_A^2, the electron density of the anti-bonding MO, and the upper solid line(dark) indicates ψ_B^2, the electron density of the bonding MO.

Fig.7.1.3 is an energy level diagram graphically indicating the energies of the two MO's that arise from the interactions of two atomic orbitals. Overlap integrals are generally small and are often in the range of 0.2 to 0.3. Hence, the anti-bonding MO is destabilized approximately the same amount that the bonding MO is stabilized.

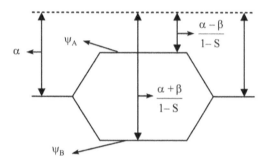

Fig.7.1.3 Energy level diagram for the molecular orbitals formed from similar atomic orbitals in a homo nuclear molecule.

In fact, in simple LCAO theory, it is often assumed that S =0. This assumption simplifies the calculations. With this assumption, the energy levels ψ_A and ψ_B are equal to $\alpha - \beta$ and $\alpha + \beta$, respectively.

In the case of a hetero-nuclear bond such as in the LiH⁺, if we neglect S, the secular determinant yields,

$$\begin{vmatrix} \alpha_1 - E & \beta - ES \\ \beta - ES & \alpha_2 - E \end{vmatrix} = 0$$

$$\begin{vmatrix} \alpha_1 - E & \beta \\ \beta & \alpha_2 - E \end{vmatrix} = 0$$

Solving $(\alpha_1 - E)(\alpha_2 - E) = \beta^2$ for E gives

$$E = \frac{\alpha_1 + \alpha_2}{2} \pm \frac{1}{2}\sqrt{\alpha_1^2 + \alpha_2^2 + 2\alpha_1\alpha_2 - 4\alpha_1\alpha_2 + 4\beta^2}$$

$$\frac{\alpha_1 + \alpha_2}{2} \pm \frac{1}{2}\sqrt{(\Delta\alpha)^2 + 4\beta^2}$$

The corresponding energy level diagram is given below.

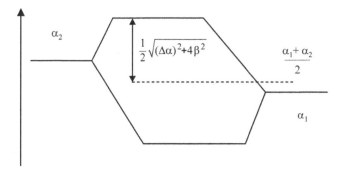

Fig.7.1.4 Energy level diagram for the molecular orbitals formed from dissimilar atomic orbitals in a hetero nuclear molecule.

In the above energy level diagram for the molecular orbitals formed from dissimilar atomic orbitals in a hetero-nuclear molecule the overlap integral has been neglected.

Notice that, according to the approximate LCAO method that we are now employing, the energy of the bonding MO is depressed from that of the more stable atomic orbital by the same amount that the energy of the anti-bonding MO is raised from that of the less stable atomic orbital. If $|\beta|$ is very small, the energy spread between the bonding and anti-bonding levels is just more than the separation between α_1 and α_2, and then the MO's are essentially slightly perturbed atomic orbitals.

Hamiltonian operator for H_2^+ and H_2

H_2^+ ion - It is represented as:

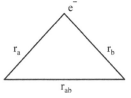

The Hamiltonian operator can be written as

$$\frac{-h^2}{8\pi^2 m}\nabla^2 + V$$

$$H = \frac{-h^2}{8\pi^2 m}\nabla^2 - \frac{e^2}{r_a} - \frac{e^2}{r_b} + \frac{e^2}{r_{ab}}$$

H_2 Molecule : The coordinates for the hydrogen molecule can be represented as

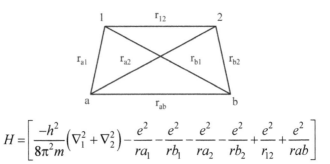

$$H = \left[\frac{-h^2}{8\pi^2 m}(\nabla_1^2 + \nabla_2^2) - \frac{e^2}{ra_1} - \frac{e^2}{rb_1} - \frac{e^2}{ra_2} - \frac{e^2}{rb_2} + \frac{e^2}{r_{12}} + \frac{e^2}{rab} \right]$$

The Stability of Hydrogen Molecule Ion

If the hydrogen molecule ion actually does form a stable species, we should expect a potential energy diagram as a function of the distance of separation of the two nuclei, to show a minimum at some equilibrium separation of the two atoms, 'a' and 'b'. It should be possible to plot such a curve, if we can evaluate the expression for the energy of the molecule as a function of the inter-nuclear separation. Immediately we should recognize that two potential energy diagrams will be obtained; one for the bonding orbital and one for the anti-bonding orbital. In both E_S and E_A the same integral will appear, but the energies will be different.

We know that

$$H_{11} = \int \phi_1 H \phi_1 d\tau$$
$$H_{22} = \int \phi_2 H \phi_2 d\tau$$
$$S_{12} = \int \phi_1 \phi_2 d\tau$$

The Hamiltonian operator for hydrogen molecule ion can be written as

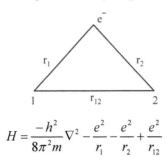

$$H = \frac{-h^2}{8\pi^2 m}\nabla^2 - \frac{e^2}{r_1} - \frac{e^2}{r_2} + \frac{e^2}{r_{12}}$$

Then

$$H_{11} = \int \phi_1 \left[\frac{-h^2}{8\pi^2 m} \nabla^2 - \frac{e^2}{r_1} - \frac{e^2}{r_2} + \frac{e^2}{r_{12}} \right] \phi_1 d\tau$$

This integral can be simplified by rearranging that

$$\frac{-h^2}{8\pi^2 m} \nabla^2 - \frac{e^2}{r_1}$$

is the Hamiltonian operator for the hydrogen atom with electron around atom-1 and since in general

$$H\phi = E\phi$$

the Hamiltonian operator for hydrogen molecule ion can be expressed as

$$H = E_0 - \frac{e^2}{r_2} + \frac{e^2}{r_{12}}$$

where E_0 is the ground state energy of the hydrogen atom. This then gives

$$H_{11} = \int \phi_1 \left[E_0 - \frac{e^2}{r_1} + \frac{e^2}{r_2} \right] \phi_1 d\tau$$

Now E_0 and r_{12} are both constants and for this reason, it is possible to remove them from under the integral sign giving

$$H_{11} = E_0 \int \phi_1 \phi_2 \, d\tau + \frac{e^2}{r_{12}} \int \phi_1 \phi_2 \, d\tau + \frac{e^2}{r_2} \int \phi_1 \phi_1 d\tau$$

Since 1S wave functions are normalized, it follows that

$$H_{11} = E_0 + \frac{e^2}{r_{12}} - J$$

where J denoted the integral

$$J = \frac{e^2}{r_{12}} \int \phi_1 \phi_2 \, d\tau.$$

The evaluation of J is not a simple matter and for that reason, it is not considered here. However, this will help to discuss the shape of the potential energy diagram in terms of its contribution to the total energy of the system.

After introducing the Hamiltonian operator, the integral becomes

$$H_{12} = \int \phi_1 \left[E_0 - \frac{e^2}{r_2} + \frac{e^2}{r_{12}} \right] \phi_2 \, d\tau$$

on expanding

$$H_{12} = E_0 \int \phi_1 \phi_1 \, d\tau + \frac{e^2}{r_{12}} \int \phi_1 \phi_2 \, d\tau - \frac{e^2}{r_2} \int \phi_1 \phi_1 \, d\tau.$$

Since $\int \phi_1 \phi_1 \, d\tau$ is defined as S_{12}, then H_{12} can be expressed as

$$H_{12} = E_0 S_{12} + \frac{e^2}{r_{ab}} S_{12} - K$$

where K denotes the integral

$$K = \frac{e^2}{r_2} \int \phi_1 \phi_1 \, d\tau.$$

Just as with the integral J, the integral K is rather difficult to evaluate, but it can be helpful to see how it will affect the energy of the molecule.

When we substitute the various integrals in the equation for energy states

$$E_S = \frac{\alpha + \beta}{1 + S} = \frac{H_{11} + H_{12}}{1 + S_{12}}$$

we obtain for the symmetric state

$$E_S = \frac{E_0 + \frac{e^2}{r_{12}} - J + E_0 S_{12} + \frac{e^2}{r_{12}} S_{12} - K}{1 + S_{12}}$$

$$E_S - E_0 = \frac{e^2}{r_{12}} - \frac{J + K}{1 + S_{12}}$$

and for the antisymmetric state

$$E_A - E_0 = \frac{e^2}{r_{12}} - \frac{J - K}{1 - S_{12}}$$

Although the arithmetic is rather complex, it is possible to evaluate the integrals J and K as a function of inter-nuclear separation of the hydrogen nuclei. This result can be shown as potential energy diagram such as the one shown in the following figure.

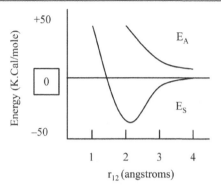

Fig. 7.1.5 Potential energy diagram showing the symmetrical and anti-symmetrical energy states.

Here the antisymmetric state is seen to correspond to an unstable energy state, and if the electron were in the anti-symmetrical orbital, we conclude that the hydrogen molecule ion would be an unstable species. On the other hand, a symmetric energy state leads to a potential minimum and therefore, a stable molecular species.

7.2 Valence Bond Theory

The problem of homo-polar bond can be seen from the point of view of Valence Bond theory. In this theory, it is assumed that atoms, complete with electrons, come together to form the molecule. The theory uses the following two principles:

(i) If ψ_A and ψ_B are wave functions for two independent systems A and B, then we can write the total wave function ψ for the separated systems as a simple product

$$\psi = \psi_A \psi_B \tag{7.2.1}$$

and the total energy $E = E_A + E_B$

(ii) If ψ_1, ψ_2, ψ_3, etc. are the acceptable wave functions for the same system, then the true wave function ψ can be obtained by taking a linear combination of all these wave functions, i.e.

$$\psi = c_1\psi_1 + c_2\psi_2 + c_3\psi_3 + \ldots\ldots \tag{7.2.2}$$

where c_1, c_2, c_3 etc. are coefficients which are adjusted to give a state of lowest energy. We can interpret the squares of the coefficients as a qualitative measure of the relative contribution of each wave function to the true wave function.

This theory was first applied by Heitler and London in 1927 to the hydrogen molecule. We shall begin with two hydrogen atoms far enough apart so that no appreciable interaction can occur. Although the two hydrogen atoms are identical, for convenience of treatment we may label the electrons as 1 and 2 and the nuclei as A and B; the orbital wave functions for the separate atoms $H_A(e_1)$ and $H_B(e_2)$ would then be given by $\psi_A(1)$ and $\psi_B(2)$, respectively. By using Eq. 7.2.1 the total orbital wave function for the separated atoms can be written as

$$\psi_N = \psi_A(1)\ \psi_B(2) \tag{7.2.3}$$

Using the wave function ψ_N the energy of the system comprising two identical hydrogen atoms can be calculated as a function of the inter nuclear distance r_{AB}. This is shown graphically in Fig.7.2.1. It may be noted that in the Fig.7.2.1 the total energy of the two isolated hydrogen atoms at infinite separation has been taken as zero, so that the energy curve on this plot shows

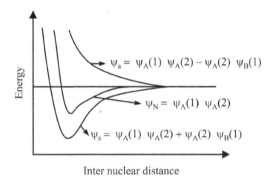

Fig.7.2.1 Variation of energy with inter-nuclear distance for different wave functions.

how much the energy of the system of two hydrogen atoms is above or below that of the two isolated atoms. Consequently, the energy value at the point of minimum on an energy curve represents the bonding energy at the equilibrium inter-nuclear distance for the molecule described by the corresponding wave function. It is seen that the energy curve N exhibits a minimum, thus indicating that a molecule is formed. However, the bonding energy is far too small, that is about 6 kcal/mole, which is only a small fraction of the observed value, namely 109 kcal/mole.

Evidently, the wave function of Eq.7.2.3 can not be correct. To reveal this error, we must recall that in forming the wave function of Eq. 7.2.3 we supposed that the two electrons were distinguishable, so that the electrons, labelled *1* and *2*, could be associated with the nuclei *A* and *B*, respectively. We know, however, that when the two atoms are very close together, so that the atomic orbitals overlap, we can no longer be sure that the electron *1* will always be near the nucleus A and the electron 2 near the nucleus B. We can, in fact, no longer distinguish one electron from the other and hence the system of two H atoms may be represented by the different states I and II as

$$H_A(e_1) \; H_B(e_2) \quad H_A(e_2) \; H_B(e_1)$$
$$\quad I \quad\quad\quad\quad\quad II$$

Let ψ_I and ψ_{II} be the wave functions for the states I and II, respectively. Following Eq. 7.2.1 they can be written as

$$\psi_I = \psi_A(1) \; \psi_B(2)$$
$$\psi_{II} = \psi_A(2) \; \psi_B(1) \tag{7.2.4}$$

The true wave function ψ, is likely to be some combination of these two wave functions. Following Eq. 7.2.2 we can write

$$\psi = c_1 \psi_I + c_2 \psi_{II}$$
$$= c_1 \psi_A + \psi_A(2) + c_2 \psi_A(2) \; \psi_B(1) \tag{7.2.5}$$

In the case of hydrogen molecule, because of symmetry, the two component wave functions ψ_I and ψ_{II} must contribute with equal weight. As the weight is proportional to the square of the coefficients we can write $c_1^2 = c_2^2$ or $c_1 = \pm c_2$. Moreover, since the coefficients are relative quantities we can put $c_1 = 1$ and hence $c_2 = \pm 1$. There are thus two possible wave functions.

$$\psi_s = \psi_A(1) \; \psi_B(2) + \psi_A(2) \; \psi_B(1) \tag{7.2.6}$$
$$\psi_a = \psi_A(1) \; \psi_B(2) - \psi_A(2) \; \psi_B(1) \tag{7.2.7}$$

Eq. 7.2.6 represents symmetric combination since ψ_s remains unchanged by exchange of electrons 1 and 2, where Eq.7.2.7 represents the antisymmetric combination since ψ_a changes sign with exchange of electrons. The energy of the system (as a function of inter-nuclear distance) calculated by using the wave function ψ_s is shown as curve A. It is thus evident from these energy curves that ψ_a represents a repulsive or nonbonding state, but ψ_s represents a bonding state. We thus find that the anti-contiguous H atoms will always lead to repulsion between the two atoms, so that the stable H_2 molecule can not be formed. While the symmetric combination will lead to attraction of the two hydrogen atoms with the formation of a stable H_2 molecule, the equilibrium value for the inter-nuclear distance is given by r_0, at which the energy is minimum. The value of r_0 found to be 0.80 Å is in rough agreement with the experimental value of 0.72 Å. The bonding energy corresponding to this separation is found to about 72 kcal/mole, which is in better agreement with the experimental value (109 kcal/mole) than the energy calculated from Eq. 7.2.3. Evidently, this improvement in result has stemmed from the introduction of the concept of electron exchange between the two constituent atoms. The additional bonding energy, which has resulted from this, Viz. (72-6) or 66 kcal/mole, is consequently known as the exchange energy.

It turns out from the above discussion that ψ_s and ψ_a are the orbital wave functions of the normal hydrogen molecule. It is also referred to as covalent wave functions because one electron is considered to be on one nucleus and another on the other.

It may be recalled that the calculated bonding energy (≈ 72 Kcal/mole), obtained by using the orbital wave function ψ_s is considerable in magnitude, it is still far short of the experimental value of about 109 Kcal/mole. The orbital wave function for the hydrogen molecule may, however, be further improved by modifying the wave functions of atomic orbitals of the electrons. The modification can be done slightly by allowing for the screening of nucleus A from nucleus B by increased probability of the electrons being found directly between the nuclei, and also to allow for the fact that the atomic orbitals will no longer be spherically symmetrical when the atoms are close together.

An additional improvement is also obtained by allowing for the possibility that both electrons may be simultaneously near one of the nuclei giving rise to the following two equally likely ionic configurations:

$$(e_2) \qquad\qquad (e_2)$$
$$H_A(e_1)H_B \quad \text{and} \quad H_AH_B(e_1)$$
$$\text{III} \qquad\qquad \text{IV}$$

The wave function for the ionic form of hydrogen molecule may therefore be represented equally well by $\psi_A(1)\psi_A(2)$ or $\psi_B(1)\psi_B(2)$. Inclusion of all these modifications leads to a bonding energy of about 95 kcal/mole. Further refinements can be effected by taking into account explicitly the inter-electron distance, and these lead to a bonding energy, which is only 0.5 kcal/mole less than the experimental value. This demonstrates strikingly the validity of the VB approach.

Taking into consideration the possibility of ionic configurations, the combined wave function for the H_2 molecule can be written as

$$\psi(s) = [\psi_A(1)\ \psi_B(2) + \psi_A(2)\ \psi_B(1)] + [\psi_A(1)\ \psi_A(2) + \psi_B(1)\ \psi_B(2)] \qquad (7.2.8)$$

or more conveniently as

$$\psi_{(s)} = \psi_{Cov} + \psi_{Ion}$$

when atoms are brought together, so that atomic orbitals each containing one electron overlap, a combined wave function of the form of Eq.7.2.8 is obtained. This, indeed, is the essential principle of Valence Bond Theory and the result is the pairing of electrons, which is equivalent to the formation of a single bond. It might be mentioned that in molecules that are more complicated multiple bonds may arise from the pairing of four or six electrons; in such cases the principle of maximum overlapping is used to decide the way in which electrons are paired.

The coefficient λ in Eq.7.2.8 is a measure of the degree to which the ionic forms contribute to the bonding. Since λ can be adjusted to give the best value of the bonding energy, it affords a means of judging the extent of ionic character of the bond. For instance, for H_2, λ is 0.17 and the ionic contribution to the total bonding energy amounts to only 5.5 kcal/mole.

The physical interpretation of Eq. 7.2.8 is that the molecule is represented adequately neither by the pure covalent structure H-H nor by the ionic structures $H_A - H_B^+$, but that the true state of affairs lies somewhere between these two extremes. In such a case, it is said that there is a resonance between the two structures. Because of resonance, the total energy of the system will seek a minimum value lying below that for any of the resonating structures. This gives rise to extra stability of the actual molecule, measured in terms of the so-called resonance energy, which is taken to be equal to the difference in energy between that of the actual molecule and the most stable of the resonating structures. This concept of resonance is of fundamental importance in Valence Bond treatment.

7.3 Hybridization

In order to explain the directional characteristics of covalent bonds in polyatomic molecules, Pauling and Slater suggested that the formation of covalent bond takes place in the direction in which there is maximum overlap. It was considered that in case of water and ammonia molecules the hydrogen atom orbitals overlap with the 'p' orbitals and hence should be at an angle of 90°. The greater bond angle in H_2O [105°] or NH_3 [108°] is supposed to be due to repulsion between partially positive hydrogen atoms. However, the calculation of columbic repulsion shows that the angle cannot deviate from 90° to 105° in water and 108° in case of NH_3.

Further, in case of BeH_2, BF_3 and CH_4 molecules the number of unpaired electrons present on the central atom is less than the number of covalent bonds formed. In order to explain this, it was presumed that the electrons are excited to the higher orbitals during the formation of covalent bonds. For example, beryllium has the excited state configuration $1s^2$, $2s^1$, $2p_x^1$. This should result in two nonequivalent bonds due to the overlap of the hydrogen atoms with 2s and $2p_x$ orbitals. However, in BeH_2 both the bonds are equivalent and are at an angle of 180°. In order to explain such cases the concept of hybridization was introduced.

According to the concept of hybridization, in cases where pure orbitals cannot affect good overlap in the formation of covalent bonds, combination of pure orbitals having same or similar energy takes place, resulting in the formation of equivalent hybrid orbitals. The number of hybrid orbitals formed is equal to the number of combining pure orbitals. If suppose ϕ_1 and ϕ_2 are the two combining atomic orbitals the resulting hybridized orbitals ψ_{h_1} and ψ_{h_2} are

$$\psi_{h_1} = c_1\phi_1 + c_2\phi_2$$
$$\psi_{h_2} = c_3\phi_1 + c_4\phi_2 .$$

The values of the coefficients should be such that each of the hybridized orbital is normalized and the two hybrid orbitals should be orthogonal to each other.

i.e. $\quad \int \psi_{h_1}^2 d\tau = \int \psi_{h_2}^2 d\tau = 1$

and $\quad \int \psi_{h_1} \psi_{h_2} d\tau = 0 .$

Hence, $\quad c_1^2 + c_2^2 = 1$ and $c_3^2 + c_4^2 = 1.$

Further, $\int (c_1\phi_1 + c_2\phi_2)(c_3\phi_1 + c_4\phi_2) d\tau = 0$

or $\quad c_1 c_3 + c_2 c_4 = 0 .$

Thus, the hybrid orbitals are orthogonal to each other. They provide better overlap with the incoming atomic orbitals and result in the formation of more stable bonds and a state of lower energy. The energy liberated is called hybridization energy. This energy is partly used for the excitation of electrons from lower to higher orbitals in the atom.

Let us now consider the compositions of different hybrid orbitals.

Linear Structure – $BeCl_2$

Here the chlorine atom p_x orbitals interact with the two hybrid orbitals directed at an angle of 180°, say along z – axis.

The 2s orbital of Be atom is spherically symmetrical and hence contributes to the formation of both the hybrid orbitals. p_z orbital is also directed in the direction in which the hybrid orbitals are formed and hence can contribute to their formation. However, p_x and p_y have nodes along z – axis and hence cannot contribute to the formation of the hybrid orbitals along that direction. Thus, the two hybrid orbitals have contributions only from s and p_z orbitals, hence they are sp hybrid orbitals. The compositions of the two hybrid orbitals can be shown as follows.

$$\psi_{h_1} = a_1 \psi_s + b_1 \psi_{p_z}$$

$$\psi_{h_2} = a_2 \psi_s + b_2 \psi_{p_z}$$

Since the s orbital is spherically symmetrical, it contributes equally to the making of the two hybrid orbitals. In terms of probability contribution it is ½ and hence the coefficient of the s – orbital wave function in making the hybrid orbitals in each case shall be, $\sqrt{1/2}$.

Hence, $\psi_{h_1} = \dfrac{1}{\sqrt{2}} \psi_s + b_1 \psi_{p_z}$

$\psi_{h_2} = \dfrac{1}{\sqrt{2}} \psi_s + b_2 \psi_{p_z}$

Since each hybrid orbital is normalized,

$$a_1^2 + b_1^2 = 1 \text{ and } a_1^2 + \dfrac{1}{2}, \text{ then}$$

$$b_1^2 = \dfrac{1}{2} \text{ and } b_1 = \sqrt{1/2}.$$

Hence, $\psi_{h_1} = \dfrac{1}{\sqrt{2}}(\psi_s + \psi_{p_z})$.

The two hybrid orbitals are orthogonal to each other.

Hence, $a_1 a_2 + b_1 b_2 = 0$

$$1/2 + 1/2 \, b_2 = 0$$

$\therefore \quad b_2 = -1/2$.

Hence $\psi_{h_2} = \dfrac{1}{\sqrt{2}}(\psi_s - \psi_{p_z})$.

The two sp hybrid orbitals are thus equivalent, one having contribution from positive phase of p_z orbital and another from the negative phase of p_z orbital. The values of s and p_z orbitals can be put in terms of wave function and it can be shown that the two hybrid orbitals formed will have maximum probability directed at an angle of 180° only and thus form more stable bonds.

Trigonal Planar Structure

The hybrid orbitals are directed to the corners of a trigonal plane. ψ_{h_1} is directed along x – axis while ψ_{h_2} and ψ_{h_3} are in between x and y axes. All the hybrid orbitals are at an angle of 120°.

The composition of the hybrid orbitals can be shown as follows:

$$\psi_{h_1} = a_1 \psi_s + b_1 \psi_{p_x} + c_1 \psi_{p_y}$$
$$\psi_{h_2} = a_2 \psi_s + b_2 \psi_{p_x} + c_2 \psi_{p_y}$$
$$\psi_{h_3} = a_3 \psi_s + b_3 \psi_{p_x} + c_3 \psi_{p_y}$$

Since s orbital is spherically symmetrical, it contributes equally to the making of the three hybrid orbitals.

Therefore, $a_1 = a_2 = a_3 = \dfrac{1}{\sqrt{3}}$.

ψ_{h_1} is formed along x – axis and hence cannot have any contribution from p_y that is $c_1 = 0$.

$$\psi_{h_1} = \left(\dfrac{1}{\sqrt{3}}\right)\psi_s + b_1 \psi_{p_x}$$

ψ_{h_1} is normalized and hence,

$$a_1^2 + b_1^2 = 1$$

$$\dfrac{1}{3} + b_1^2 = 1; \therefore b_1^2 = \dfrac{2}{3} \text{ and } b_1 = \sqrt{\dfrac{2}{3}}.$$

$$\psi_{h_1} = \dfrac{1}{\sqrt{3}}\psi_s + \sqrt{\dfrac{2}{\sqrt{3}}}\psi_{p_x}.$$

Now considering that ψ_{h_1} and ψ_{h_2} are orthogonal to each other,

$$a_1 a_2 + b_1 b_2 = 0 \qquad (c_1 c_2 = 0)$$

$$\dfrac{1}{3} + \sqrt{\dfrac{2}{3}} b_2 = 0$$

$$\therefore b_2 = \dfrac{-1}{\sqrt{6}}.$$

The normalization condition requires

$$a_2^2 + b_2^2 + c_2^2 = 1$$

$$\therefore \frac{1}{3} + \frac{1}{6} + c_2^2 = 1, \quad \therefore c_2^2 = \frac{1}{2}$$

So $c_2 = \dfrac{1}{\sqrt{2}}$

Hence, $\psi_{h_2} = \dfrac{1}{\sqrt{3}} \psi_s - \dfrac{1}{\sqrt{6}} \psi_{px} + \dfrac{1}{\sqrt{2}} \psi_{py}$

Considering the orthogonality of ψ_{h_1} and ψ_{h_3} it can be shown that

$$a_1 a_3 + b_1 b_3 = 0, \text{ hence } b_3 = -\frac{1}{\sqrt{6}}.$$

The orthogonality condition of ψ_{h_2} and ψ_{h_3} requires

$$a_2 a_3 + b_2 b_3 + c_2 c_3 = 0$$

$$\frac{1}{3} + \frac{1}{6} + \frac{1}{\sqrt{2}} c_3 = 0$$

$$\frac{1}{\sqrt{2}} c_3 = -\frac{1}{2}, \quad \therefore c_3 = -\frac{1}{\sqrt{2}}.$$

Hence, $\psi_{h_3} = \dfrac{1}{\sqrt{3}} \psi_s - \dfrac{1}{\sqrt{6}} \psi_{px} + \dfrac{1}{\sqrt{2}} \psi_{py}$.

From physical considerations, the meaning of signs of coefficients can be understood. ψ_{h_1} is formed by contribution from s orbital and positive phase of p_x orbital. ψ_{h_2} has contribution from s orbital and positive phase of p_y orbital and negative phase of p_x, ψ_{h_3} has contribution from s orbital, negative phase of p_x, and negative phase of p_y orbital.

It can be seen that all the three hybrid orbitals are orthogonal. By substituting the wave functions of ψ_s, ψ_{px} and ψ_{py}, it can be shown that the hybrid orbitals are at an angle of 120° and will have greater overlap with the incoming orbitals than the corresponding pure atomic orbitals.

Tetrahedral Structure

For example in the case of methane molecule 2s, and 2p orbitals are close in energy and hence sp³ hybridization takes place. In cases where s and d orbitals are close in energy, sd³ hybridization is possible.

The four hybrid orbitals can be shown as follows:

$$\psi_{h1} = a_1\psi_s + b_1\psi_{px} + c_1\psi_{py} + d_1\psi_{pz}$$
$$\psi_{h2} = a_2\psi_s + b_2\psi_{px} + c_2\psi_{py} + d_2\psi_{pz}$$
$$\psi_{h3} = a_3\psi_s + b_3\psi_{px} + c_3\psi_{py} + d_3\psi_{pz}$$
$$\psi_{h4} = a_4\psi_s + b_4\psi_{px} + c_4\psi_{py} + d_4\psi_{pz}$$

The coefficients can be worked out by performing the projection operations $\hat{P}A_1$ and $\hat{P}T_2$ on the pendent sigma orbitals and then finding the inverse of the matrix. It is found to be as follows:

$$\begin{bmatrix}\psi_s \\ \psi_{px} \\ \psi_{py} \\ \psi_{pz}\end{bmatrix} = \begin{bmatrix} \frac{1}{2} & \frac{1}{2} & \frac{1}{2} & \frac{1}{2} \\ \frac{1}{2} & \frac{-1}{2} & \frac{-1}{2} & \frac{1}{2} \\ \frac{1}{2} & \frac{-1}{2} & \frac{1}{2} & \frac{-1}{2} \\ \frac{1}{2} & \frac{1}{2} & \frac{-1}{2} & \frac{-1}{2} \end{bmatrix} \begin{bmatrix}\psi_{h1} \\ \psi_{h2} \\ \psi_{h3} \\ \psi_{h4}\end{bmatrix}$$

$$\psi_{h1} = \frac{1}{2}(\psi_s + \psi_{px} + \psi_{py} + \psi_{pz})$$
$$\psi_{h2} = \frac{1}{2}(\psi_s - \psi_{px} - \psi_{py} + \psi_{pz})$$
$$\psi_{h3} = \frac{1}{2}(\psi_s - \psi_{px} + \psi_{py} - \psi_{pz})$$
$$\psi_{h4} = \frac{1}{2}(\psi_s + \psi_{px} - \psi_{py} - \psi_{pz})$$

The physical meaning of the coefficients can be appreciated if we consider the four hybrid orbitals to be directed to the opposite corners of the two opposite faces of the cube. s orbital is spherically symmetrical and hence contributes equally to the making of all the four hybrid orbitals. The hybrid orbitals are uniformly disposed with respect to p_x, p_y and p_z orbitals and hence they also contribute equally to the making of the hybrid orbitals. Thus the contribution is $\frac{1}{4}$ by each of the component orbitals and hence coefficient is $\frac{1}{\sqrt{4}} = \frac{1}{2}$.

Ψ_{h_1} is formed from the positive phase of all the three p orbitals. Ψ_{h_2} from -ve p_x, -ve p_y and +ve p_z, Ψ_{h_3} from -ve p_x, +ve p_y and -ve p_z and Ψ_{h_4} +ve p_x, -ve p_y and -ve p_z. Accordingly, the signs of coefficients are obtained.

All the four hybrid orbitals are orthogonal. By substituting the wave functions of the s and p orbitals and maximizing the probability, it can be seen that the hybrid orbitals have maximum probability of occurring at tetrahedral angles.

Octahedral Complexes

In octahedral case the six hybrid orbitals are formed by the combination of d_{z^2}, $d_{x^2-y^2}$, s, p_x, p_y and p_z orbitals i.e. d^2sp^3 hybridization. The orbitals involved are normally $3d_{x^2-y^2}$, $3d_{z^2}$, 4s, $4p_x$, 4py and $4p_z$ in cases of the complexes of the first transition series metal ions. This is called inner orbital hybridization. In some cases sp^3d^2 takes place and is called the outer orbital hybridization.

The coefficients of the s, p and d orbitals contributing to the six hybrid orbitals can be worked out by performing the projection operation $\widehat{PA}_{1g}$, $\widehat{PE}_g$ and $\widehat{PT}_{1u}$ on the six pendent sigma ligand orbitals and then finding the inverse of the matrix.

$$\Psi_{h_1} = \tfrac{1}{\sqrt{6}}\Psi_s + \tfrac{1}{\sqrt{2}}\Psi_{p_x} + \tfrac{1}{2}\Psi_{d_{x^2-y^2}} - \tfrac{1}{\sqrt{12}}\Psi_{d_{z^2}}$$

$$\Psi_{h_2} = \tfrac{1}{\sqrt{6}}\Psi_s + \tfrac{1}{\sqrt{2}}\Psi_{p_z} + \tfrac{1}{\sqrt{3}}\Psi_{d_{z^2}}$$

$$\Psi_{h_3} = \tfrac{1}{\sqrt{6}}\Psi_s - \tfrac{1}{\sqrt{2}}\Psi_{p_x} + \tfrac{1}{2}\Psi_{d_{x^2-y^2}} - \tfrac{1}{\sqrt{12}}\Psi_{d_{z^2}}$$

$$\Psi_{h_4} = \tfrac{1}{\sqrt{6}}\Psi_s - \tfrac{1}{\sqrt{2}}\Psi_{p_z} + \tfrac{1}{\sqrt{3}}\Psi_{d_{z^2}}$$

$$\Psi_{h_5} = \tfrac{1}{\sqrt{6}}\Psi_s + \tfrac{1}{\sqrt{2}}\Psi_{p_y} - \tfrac{1}{2}\Psi_{d_{x^2-y^2}} - \tfrac{1}{\sqrt{12}}\Psi_{d_{z^2}}$$

$$\Psi_{h_6} = \tfrac{1}{\sqrt{6}}\Psi_s - \tfrac{1}{\sqrt{2}}\Psi_{p_y} - \tfrac{1}{2}\Psi_{d_{x^2-y^2}} - \tfrac{1}{\sqrt{12}}\Psi_{d_{z^2}}$$

The physical significance of the coefficients can also be understood. The s orbital is spherically symmetrical and hence contributes equally to all the six orbitals(1/6); p_x, p_y and p_z contribute equally(1/2) only to the two hybrid orbitals along that axis in positive or negative phase. $d_{x^2-y^2}$ orbital contributes equally only to the four hybrid orbitals in the XY plane(1/4), +ve to the hybrid orbitals along the X axis and –ve to the orbitals along Y axis. d_{z^2} has greater +ve contribution(1/3) to each of the hybrid orbitals along Z axis and lesser –ve contribution(1/12) each to the four hybrid orbitals along XY plane. All the six hybrid orbitals are orthogonal.

The resulting hybrid orbital wave functions are called Symmetry Adapted Linear Combinations(SALC's). By applying the reduction formula to the total character of the hybrid orbitals, we know that the SALC's will be corresponding to the particular irreducible representations. For example A_1' and E_1' in case of sp^2 hybridization. The composition of the SALC's can be worked out by the method of projection operators.

The projection operators of A_1' irreducible representation and E_1' irreducible representation can be performed to get the proper SALC's which form the basis for A_1' and E_1' representations. The method is applied in the following way.

The equation correlating the pure atomic orbitals, the coefficients and the hybrid orbitals can be shown in the form of matrix.

$$\begin{bmatrix} \psi_{h_1} \\ \psi_{h_2} \\ \psi_{h_3} \end{bmatrix} = \begin{bmatrix} a_1 & b_1 & c_1 \\ a_2 & b_2 & c_2 \\ a_3 & b_3 & c_3 \end{bmatrix} \begin{bmatrix} \psi_s \\ \psi_{p_x} \\ \psi_{p_y} \end{bmatrix}$$

It is difficult to work out a, b, c directly because projection operator can be applied only on equivalent wave functions (atomic wave functions are not equivalent). The inverse transformation of the above matrix is therefore carried out.

$$\begin{bmatrix} \psi_s \\ \psi_{p_x} \\ \psi_{p_y} \end{bmatrix} = \begin{bmatrix} x_1 & y_1 & z_1 \\ x_2 & y_2 & z_2 \\ x_3 & y_3 & z_3 \end{bmatrix} \begin{bmatrix} \psi_{h_1} \\ \psi_{h_2} \\ \psi_{h_3} \end{bmatrix}$$

The x, y, z matrix is the inverse of a, b, c matrix. Hence x, y, z matrix can be obtained and from this a, b, c matrix can be worked out and thus, the coefficients of the pure atomic orbitals can be obtained.

The x, y, z matrix describes the transformation of a set of three equivalent basic functions(hybrid orbitals) into a set of linear combinations having the symmetry of the atomic orbitals, which in turn have symmetry corresponding to one irreducible representation of the molecular point group.

CHAPTER 8

Appendix

8.1 SI Units (Système Internationale d'unités)

When making measurements of a physical quantity, the result is expressed as a number followed by the unit. The number expresses the ratio of the measured quantity to some fixed standard and the unit is the name or the symbol for the standard.

Over the years, a large number of standards have been defined for physical measurements and many systems of units have evolved. Ex. 1. CGS, 2. FPS and 3. MKS (Metric system). Recently, there has been an attempt to simplify the language of science by the adoption of a system of units "Système Internationale d'unites", called the SI units.

SI contains three classes of units 1. Base units, 2. Derived units and 3. Supplementary units.

Base units:

Quantity	Name	Symbol
length,	meter:	m
mass	kilogram:	kg
time	second:	s
electric current	ampere:	A
temperature	kelvin:	K
amount of substance	mole:	mol
luminous intensity	candela:	cd

8.2 Derived Units

Frequency	hertz:	$Hz = 1/s$
Force	newton:	$N = m\ kg/s^2$
Pressure, stress	pascal:	$Pa = N/m^2 = kg/m\ s^2$
Energy, work, quantity of heat	joule:	$J = N\ m = m^2\ kg/s^2$

Quantity	Unit	Symbol/Definition
Power, radiant flux	watt:	$W = J/s = m^2\ kg/s^3$
Quantity of electricity, Electric charge	coulomb:	$C = s\ A$
Electric potential	volt:	$V = W/A = m^2\ kg/s^3\ A$
Capacitance	farad:	$F = C/V = s^4\ A^2/m^2\ kg$
Electric resistance	ohm:	$\Omega = V/A = m^2\ kg/s^3\ A^2$
Conductance	siemens:	$S = A/V = s^3\ A^2/m^2\ kg$
Magnetic flux	weber:	$Wb = V\ s = m^2\ kg/s^2\ A$
Magnetic flux density, Magnetic induction	tesla:	$T = Wb/m^2 = kg/s^2\ A$
Inductance	henry:	$H = Wb/A = m^2\ kg/s^2\ A^2$
Luminous flux	lumen:	$lm = cd\ sr$
Illuminance	lux:	$lx = lm/m^2 = cd\ sr/m^2$
Activity (ionizing radiations)	becquerel:	$Bq = 1/s$
Surface tension	newton per meter:	$N/m = kg/s^2$
Heat flux density,	watt per square meter:	$W/m^2 = kg/s^3$
Heat capacity, entropy	joule per kelvin:	$J/K = m^2\ kg/s^2\ K$
Specific heat capacity, specific entropy	joule per kilogram kelvin:	$J/kg\ K = m^2/s^2\ K$
Specific energy	joule per kilogram:	$J/kg = m^2/s^2$
Thermal conductivity	watt per meter kelvin:	$W/m\ K = m\ kg/s^3\ K$
Electric field strength	volt per meter:	$V/m = m\ kg/s^3\ A$
Permittivity	farad per meter:	$F/m = s^4\ A^2/m^3\ kg$
Permeability	henry per meter:	$H/m = m\ kg/s^2\ A^2$
Molar energy	joule per mole:	$J/mol = m^2\ kg/s^2\ mol$
Molar entropy,	joule per mole kelvin:	$J/mol\ K = m^2\ kg/s^2\ K\ mol$

8.3 Supplementary Units

The Radian(rad): Radian is the plane angle between two radii of a circle which cut off on the circumference an arc equal in length to the radius.

The Sterdian(sr): The Sterdian is the solid angle which, having its vertex at the center of the sphere, cuts off an area of the surface of the sphere equal to that of a square with sides of length equal to the radius of the sphere.

8.4 CGS Units

erg	1 erg = 10^{-7} J
dyne	1 dyn = 10^{-5} N
poise	1 P = 1 dyn s/cm^2 = 0.1 Pa s
stokes	1 St = 1 cm^2/s = 10^{-4} m^2/s
gauss	1 G = 10^{-4} T
oersted	1 Oe = (1000/(4 pi)) A/m
maxwell	1 Mx = 10^{-8} Wb
stilb	1 sb = 1 cd/cm^2 = 10^4 cd/m^2
phot	1 ph = 10^4 lx

8.5 Prefix Dictionary

Exponent (base 10) of decimal numbers: E n = 10^n

Factor		Prefix	Symbol
10^{24}	E 24	yotta	Y
10^{21}	E 21	zetta	Z
10^{18}	E 18	exa	E
10^{15}	E 15	peta	P
10^{12}	E 12	tera	T
10^{9}	E 9	giga	G
10^{6}	E 6	mega	M
10^{3}	E 3	kilo	k
10^{2}	E 2	hecto	h
10^{1}	E 1	deca	da
10^{-1}	E −1	deci	d
10^{-2}	E -2	centi	c
10^{-3}	E −3	milli	m
10^{-6}	E −6	micro	μ

Factor		Prefix	Symbol
10^{-9}	E–9	nano	n
10^{-12}	E–12	pico	p
10^{-15}	E–15	femto	f
10^{-18}	E–18	atto	a
10^{-21}	E–21	zepto	z
10^{-24}	E–24	yocto	y

8.6 Experimental Foundation

Energy units: The international system of units (S.I) expresses fundamental physical quantities such as mass, time, length, thermodynamic temperature and amount of substance in terms of the units kilogram (Kg), second (S), meter (M), Kelvin (K) and mole (mol) respectively.

Energy is expressed in Joules (J). Many chemists have been brought up in C.G.S. system and have been accustomed to expressing (thermochemical) energy in calories or kilocalories, which are not S.I. units. The appropriate conversion factor is,

$$1 \text{ Calorie} = 4.184 \text{ J}.$$

The energies of electrons in atoms can very conveniently be expressed in electron-volts (eV), where 1 eV is the energy acquired by an electron when it is accelerated by a potential difference of one volt.

$$1 \text{ eV} = 1.6021 \times 10^{-19} \text{ J/atom}.$$

In radiation theory, wavelength λ and frequency v (are related by $v = \dfrac{c}{\lambda}$ (where c = velocity of electromagnetic radiation in vacuum (2.9979×10^8 m/s). The frequency unit, 'v' is called the Hertz (Hz). The wavenumber, $\bar{v}$ is often used in spectroscopy, $\bar{v} = \dfrac{1}{\lambda}$ and the wavenumber unit is thus the reciprocal meter, m⁻¹. In practice, chemists normally find it more convenient to use the reciprocal centimeter, cm⁻¹.

The wavenumber is related to a frequency by $v = \dfrac{c}{\lambda}$ or $\bar{v} = \dfrac{1}{\lambda}$. Substituting the value of λ; $v = c\bar{v}$ i.e., v (Hz) = (ms⁻¹) ×(m⁻¹).

We can now use the Planck's expression for the quantum of energy, $E = hv = hc\bar{v}$.

$$E = 6.6256 \times 10^{-34} \text{JS} \times 2.9979 \times 10^8 \text{ ms}^{-1} \times (\text{m}^{-1}),$$
$$= 19.863 \times 10^{-26} \text{ J/atom}.$$

These values relate to single atoms and chemists usually refer to energy changes per mole of substance, where one mole is the amount of substance that contains as many elementary particles (electrons, atoms, molecules etc.) as there are atoms in 0.012 Kg of carbon-12. This number is the Avogadro number, N where, N = 6.0225 × 10^{23} atoms mole^{-1}.

Then the relationship between energy and wave number becomes,

$$E(J\ mole^{-1}) = 19.863 \times 10^{-26} \times 6.0225 \times 10^{23}\ \bar{v}$$

$$E = h\bar{v}c = JS \times ms^{-1} \times m^{-1}$$

$$E = 1.986 \times 10^{-25}\ \bar{v}\ J/atom \times 6.022 \times 10^{23}\ atom/mol$$

$$= 0.1196\ \bar{v}\ J\ mole^{-1}$$

and $\quad E(eV) = 1.6021 \times 10^{-19} J/atom \times 6.0225 \times 10^{23}\ atom/mol$

$$= 9.649 \times 10^{4}\ J\ mol^{-1}.$$

8.7 Calculation of Effective Nuclear Charge

In the case of an atom (atomic number Z) consisting of a positive nucleus (charge Ze) surrounded by 'Z' electrons, a given electron 'i' will be subjected not only to the attractive potential field of the nucleus as in the case of a single electron of the hydrogen atom, but also to the repulsive potential due to all other electrons.

So, there are two opposing factors which have to be accounted for.
1. Attraction between the positive nucleus and the electron under consideration.
2. Repulsion due to the negatively charged electrons with the electron under consideration.

Therefore, the net result will be the electron under consideration experiences less nuclear attraction because of the presence of other electrons. In other words, the electron under consideration is said to be screened from the nucleus.

Hence, the potential $V_{(i)} = \dfrac{-Ze^2}{r_i} + \sum \dfrac{e^2}{r_{ij}}$ where 'r_i' is the distance of the 'ith' electron from the nucleus of charge +Ze and r_{ij} is the distance between the 'ith' and the 'jth' electrons.

However, it is reasonable to replace the above potential by an effective potential V(r$_i$) for the 'ith' electron, which involves only 'r_i; and is termed a central potential.

This central potential is used for solving the Schrodinger equation for complex atom. This method is called self-consistent field method (Hartee-Fock method).

The wave function can be expressed algebraically as a sum of simple hydrogenic radial functions of the form,

$$\Psi_{\alpha(r,\theta,\phi)} = \sum_{i=1}^{i=p} C_{n_i} r^{(n^*-1)} \exp^{(\alpha,r)} Y_{l\pm m}^{(\theta,\phi)}$$

where C_n is the numerical coefficient and the integer 'p' depends on the extent of the matching of the numerical functions. Slater (1930) proposed a single parameter functions to represent the above sum and proposed a set of rules to determine 'n' and the orbital exponent α_i.

n* is calculated for the corresponding real quantum number 'n' as:

n = 1.0, 2.0, 3.0, 4.0, 5.0, 6.0
n* = 1.0, 2.0, 3.0, 3.7, 4.0, 4.2

The carbon electronic configuration is $1s^2\ 2s^2\ 2p^2$. Hence, the various Slater radial functions are,

Ex:
$$R_{1s} = N_{1s} \exp\left[\frac{\alpha_{1s} r}{a_0}\right]$$

$$R_{2s} = N_{2s} \exp\left[\frac{\alpha_{2s} r}{a_0}\right]$$

$$R_{2p} = N_{2p} \exp\left[\frac{\alpha_{2p} r}{a_0}\right]$$

where N is normalizing constant and $a_0 = \frac{\hbar^2}{me^2}$, $\alpha = \frac{Z-S}{n^*}$,

Z = atomic number; S = screening constant,

Ex. For $\alpha_{1s} = \frac{6 - 0.30}{1} = 5.7$

(carbon Z = 6 and S = 0.3).

This value is used to calculate R_{1s}.

The effective nuclear charge Z* acting on the electron is given by Z* = Z - S, where Z is the atomic number and S is a screening constant.

To determine S, divide the electrons into the respective orbital groups.

1s,
2s, 2p,
3s, 3p,

3d,
4s, 4p,
4d,
4f,
5s, 5p,

The 'S' is the sum of the contributions of:
(a) Zero from any orbital group outside the one considered.
(b) 0.35 in general, but 0.30 in the case of 1s, from every other electron in the orbital group considered.
(c) 0.85 from every electron in the quantum level immediately below (near to the nucleus) than the electron considered, and 1.00 from every electron in levels still nearer the nucleus, provided that the electron considered is in an 's' or a 'p' orbital.
(d) If the electron considered is in a 'd' or 'f' orbital, every electron in lower orbital groups contributed 1 towards the value of 'S'.

Thus for an electron in the 3s or 3p shell in silicon $1s^2$, $2s^2$, $2p^6$, $3s^2$, $3p^2$ (Z = 14)

$$S = (3 \times 0.35) + (8 \times 0.85) + (2 \times 1) = 9.85$$
$$Z* = 14 - 9.85 = 4.15.$$

1. **Calculate the effective nuclear charge for the following:,**
   1. He: $1s2$: $Z = Z - S = 2 - (0.30 \times 1) = 1.7$.
   2. O : $1s2\ 2s2\ 2p4$,
      $$Z* = 8 - [(0.35 \times 5) + (0.85) \times 2]$$
      $$= 8 - 1.75 - 1.70 = 4.55$$
   3. Cl–: $1s2\ 2s2\ 2p6\ 3s2\ 3p6$,
      $$Z* = 17 - (0.35 \times 7) - (0.85 \times 8) - (2 \times 1) = 5.75$$
   4. K+: $1s2\ 2s2\ 2p6\ 3s2\ 3p6$,
      $$Z* = 19 - (0.35 \times 7) - (0.85 \times 8) - (2 \times 1) = 7.75.$$
   5. Ga: $1s2\ 2s2\ 2p6\ 3s2\ 3p6\ 3d10\ 4s2\ 4p1$,
      $$Z* = 31 - (0.35 \times 2) - (0.85 \times 18) - (10 \times 1) = 5.$$
   6. Mn2+: $1s2\ 2s2\ 2p6\ 3s2\ 3p6\ 3d5$,
      $$Z* = 25 - (0.35 \times 4) - (18 \times 1) = 5.6.$$

2. **Calculate the radii of K^+ and Cl^- ions in KCl given that the bond length of KCl is 3.14 Å.**

   It is known that the radius of an ion of an atom is inversely proportional to its effective nuclear charge operative on its outermost electrons.

From the previous problems,

$$Z^*_{Cl^-} = 5.75 \text{ and } Z^*_{K^+} = 7.75,$$

$$\frac{r_{K^+}}{r_{Cl^-}} = \frac{Z^*_{Cl^-}}{Z^*_{K^+}} = \frac{5.75}{7.75} \text{ we also have } r_{K^+} + r_{Cl^-} = 3.14 \text{ Å}$$

∴ $r_{Cl^-} = 1.803$ Å and $r_{K^+} = 1.337$ Å

8.8 Approximate Orbitals

If an atom has more than one electron, we are forced to develop ways of finding approximate solutions of the Schrödinger equation. The simplest approximation is to ignore the influence one has on the other. However, this type of neglect is not justified. There are a number of facets to electron correlation.

1. The first, and the most obvious is that owing to their similar charges electrons will avoid being in the same region of space. That is, they will tend to avoid being at the same distance along a radius. This is "radial correlation".
2. They will also avoid being at the same angle to the nucleus. Indeed, all other things being equal, they will tend to be found on opposite sides of the nucleus. This is "angular correlation".
3. Of greater subtlety is the "spin correlation", which has its explanation in the Pauli's exclusion principle. For reasons nothing to do with their charge, electrons with the same spin are unlikely to be found in the same region of space.

A responsible appropriate wave function for an atom should take account of the three types of correlations. However, to do so requires a great deal of integrity and effort. Fortunately, for many purposes, we can derive much valuable information from the use of wave functions, which at first sight appear to be quite crude.

Slater orbitals

In 1930 J.C. Slater proposed a set of rules for taking into account the influence of shielding. The angular wave functions derived from the exact solution of the Schrödinger equation were preserved, but the radial wave functions were replaced by a new set.

We shall write the Slater radial wave functions as

$$R_S(r) = N_n r^{n-1} e^{\frac{-\zeta r}{a_0}}$$

N_n = normalization constant and $\varsigma = \dfrac{Z^*}{n^*}$ (original hydrogen atom wave functions used are $\dfrac{Z}{n}$.

n	1	2	3	4	5	6
n*	1	2	3	3.7	4.0	4.2

Here Z^* is the effective nuclear charge (See appendix 8.7) and 'n*' is the corresponding value of 'n' in hydrogen wave functions.

Differences between Hydrogen like and Slater orbitals

1. The Slater type orbitals(STO) ignore all but the higher power of 'r'. This means that STO's are not very good approximations close to the nucleus, but they improve as 'r' increases.

   Therefore, they give better predictions of ionization energies compared to x-ray spectra.

2. The second difference is the replacement of the exponential factor $e^{\dfrac{-zr}{na_0}}$ by the Slater factor $e^{\dfrac{-\varsigma r}{a_0}}$ (where $\varsigma = \dfrac{Z_{eff}}{n}$).

8.9 Angular Momentum

Angular momentum is an important dynamical variable. For a single particle moving around a fixed point, the angular momentum L is given by the product of 'r' and 'p' can be written in terms of their components as

$$r = ix + jy + kz$$
$$p = ip_x + jp_y + kp_z$$

where i, j and k are unit vectors along x, y and z axis. Therefore, in terms of the components of 'r' and 'p', the angular momentum, L is

$$L = r \times p = (ix + jy + kz)(ip_x + jp_y + kp_z)$$
$$= i(yp_z - zp_y) + j(zp_x - xp_z) + k(xp_y - yp_x)$$

Replacing p's by the corresponding quantum mechanical operators, the operators for the components for the angular momentum are as follows:

$$\hat{L}_x = -\dfrac{ih}{2\pi}(y\dfrac{\partial}{\partial z} - z\dfrac{\partial}{\partial y})$$

148   Quantum Chemistry

$$\hat{L}_y = -\frac{ih}{2\pi}(z\frac{\partial}{\partial x} - x\frac{\partial}{\partial z})$$

$$\hat{L}_z = -\frac{ih}{2\pi}(x\frac{\partial}{\partial y} - z\frac{\partial}{\partial x})$$

The total angular momentum is obviously given by

$$L = iL_x + jL_y + kL_z$$

However, more important in quantum mechanics is the scalar product of 'L' with itself.

$$L.L = L_x^2 + L_y^2 + L_z^2.$$

The angular momentum operators are usually expressed in spherical polar coordinates.

$$x = r\sin\theta\cos\phi \qquad z = r\cos\theta$$
$$y = r\sin\theta\sin\phi \quad \text{and} \quad x^2 + y^2 + z^2 = r^2$$
$$z = r\cos\theta \qquad \cos\theta = z\Big/\sqrt{(x^2 + y^2 + z^2)}$$

$$\psi = f(r,\theta,\phi)$$

$$\partial\psi = \frac{\partial\psi}{\partial r}\partial r + \frac{\partial\psi}{\partial\theta}\partial\theta + \frac{\partial\psi}{\partial\phi}\partial\phi$$

$$\frac{\partial}{\partial x} = \frac{\partial}{\partial r}\frac{\partial r}{\partial x} + \frac{\partial}{\partial\theta}\frac{\partial\theta}{\partial x} + \frac{\partial}{\partial\phi}\frac{\partial\phi}{\partial x}$$

$$\frac{\partial}{\partial y} = \frac{\partial}{\partial r}\frac{\partial r}{\partial y} + \frac{\partial}{\partial\theta}\frac{\partial\theta}{\partial y} + \frac{\partial}{\partial\phi}\frac{\partial\phi}{\partial y}$$

$$\frac{\partial}{\partial z} = \frac{\partial}{\partial r}\frac{\partial r}{\partial z} + \frac{\partial}{\partial\theta}\frac{\partial\theta}{\partial z} + \frac{\partial}{\partial\phi}\frac{\partial\phi}{\partial z}$$

By differentiating x, y, z with respect to r, θ and ϕ, we get

$$\frac{\partial r}{\partial x} = \sin\theta\cos\phi \qquad \frac{\partial r}{\partial y} = \sin\theta\sin\phi \qquad \frac{\partial r}{\partial z} = \cos\theta$$

$$\frac{\partial\theta}{\partial x} = \frac{\cos\theta\cos\phi}{r} \qquad \frac{\partial\theta}{\partial y} = \frac{\cos\theta\sin\phi}{r} \qquad \frac{\partial\theta}{\partial z} = -\frac{\sin\theta}{r}$$

$$\frac{\partial\phi}{\partial x} = -\frac{\sin\phi}{r\sin\theta} \qquad \frac{\partial\phi}{\partial y} = \frac{\cos\phi}{r\sin\theta} \qquad \frac{\partial\phi}{\partial y} = 0$$

Therefore,

$$\hat{L}_x = -\frac{ih}{2\pi}(y\frac{\partial}{\partial z} - z\frac{\partial}{\partial y})$$

$$= -\frac{ih}{2\pi}\begin{bmatrix} r\sin\theta\sin\phi(\cos\theta\frac{\partial}{\partial r} - \frac{\sin\theta}{r}\frac{\partial}{\partial\theta}) - r\cos\theta(\sin\theta\sin\phi\frac{\partial}{\partial r} \\ + \frac{\cos\theta\sin\phi}{r}\frac{\partial}{\partial\theta} + \frac{\cos\phi}{r\sin\theta}\frac{\partial}{\partial\phi} \end{bmatrix}$$

$$= -\frac{ih}{2\pi}\begin{bmatrix} r\sin\theta\sin\phi\cos\theta\frac{\partial}{\partial r} - \sin^2\theta\sin\phi\frac{\partial}{\partial\theta} - r\cos\theta\sin\theta\sin\phi\frac{\partial}{\partial r} \\ -\cos^2\theta\sin\phi\frac{\partial}{\partial\theta} - \frac{\cos\theta\cos\phi}{\sin\theta}\frac{\partial}{\partial\phi} \end{bmatrix}$$

$$= -\frac{ih}{2\pi}\left[-\sin\phi\frac{\partial}{\partial\theta} - \cot\theta\cos\phi\frac{\partial}{\partial\phi}\right]$$

$$\hat{L}_y = -\frac{ih}{2\pi}(z\frac{\partial}{\partial x} - x\frac{\partial}{\partial z})$$

$$= -\frac{ih}{2\pi}\begin{bmatrix} r\cos\theta(\sin\theta\cos\phi\frac{\partial}{\partial r} + \frac{\cos\theta\cos\phi}{r}\frac{\partial}{\partial\theta} - \frac{\sin\phi}{r\sin\theta}\frac{\partial}{\partial\phi}) \\ -r\sin\theta\cos\phi(\cos\theta\frac{\partial}{\partial r} - \frac{\sin\theta}{r}\frac{\partial}{\partial\theta}) \end{bmatrix}$$

$$= -\frac{ih}{2\pi}\begin{bmatrix} r\cos\theta\sin\theta\cos\phi\frac{\partial}{\partial r} + \cos^2\theta\cos\phi\frac{\partial}{\partial\theta} - \frac{\cos\theta\sin\phi}{\sin\theta}\frac{\partial}{\partial\phi} \\ -r\sin\theta\cos\phi\cos\theta\frac{\partial}{\partial r} + \sin^2\theta\cos\phi\frac{\partial}{\partial\theta} \end{bmatrix}$$

$$= -\frac{ih}{2\pi}\left[\cos\phi\frac{\partial}{\partial\theta} - \cot\theta\sin\phi\frac{\partial}{\partial\phi}\right]$$

$$\hat{L}_z = -\frac{ih}{2\pi}(x\frac{\partial}{\partial y} - z\frac{\partial}{\partial x})$$

$$= -\frac{ih}{2\pi}\begin{bmatrix} r\sin\theta\cos\phi(\sin\theta\sin\phi\frac{\partial}{\partial r} + \frac{\cos\theta\sin\phi}{r}\frac{\partial}{\partial\theta} + \frac{\cos\phi}{r\sin\theta}\frac{\partial}{\partial\phi}) \\ -r\sin\theta\sin\phi(\sin\theta\cos\phi\frac{\partial}{\partial r} + \frac{\cos\theta\cos\phi}{r}\frac{\partial}{\partial\theta} - \frac{\sin\phi}{r\sin\theta}\frac{\partial}{\partial\phi}) \end{bmatrix}$$

$$= -\frac{ih}{2\pi}\begin{bmatrix} r\sin^2\theta\cos\phi\sin\phi\frac{\partial}{\partial r} + \sin\theta\cos\theta\cos\phi\sin\phi\frac{\partial}{\partial\theta} + \cos^2\phi\frac{\partial}{\partial\phi} \\ -r\sin^2\theta\cos\phi\sin\phi\frac{\partial}{\partial r} - \sin\theta\cos\theta\cos\phi\sin\phi\frac{\partial}{\partial\theta} - \sin^2\phi\frac{\partial}{\partial\phi} \end{bmatrix}$$

$$= -\frac{ih}{2\pi}\frac{\partial}{\partial\phi}$$

$$\hat{L}^2 = -\frac{h^2}{4\pi^2}\left[\frac{1}{\sin\theta}\frac{\partial}{\partial\theta}(\sin\theta\frac{\partial}{\partial\theta}) + \frac{1}{\sin^2\theta}\frac{\partial^2}{\partial\phi^2}\right]$$

8.10 Laplacian Operator
(Conversion from Cartesian to Polar coordinates)

$$\nabla^2\psi = \frac{\partial^2\psi}{\partial x^2} + \frac{\partial^2\psi}{\partial y^2} + \frac{\partial^2\psi}{\partial z^2}$$

$$\frac{\partial\psi}{\partial x} = \frac{\partial\psi}{\partial r}\frac{\partial r}{\partial x} + \frac{\partial\psi}{\partial\theta}\frac{\partial\theta}{\partial x} + \frac{\partial\psi}{\partial\phi}\frac{\partial\phi}{\partial x}$$

$$= \psi'_r\frac{\partial r}{\partial x} + \psi'_\theta\frac{\partial\theta}{\partial x} + \psi'_\phi\frac{\partial\phi}{\partial x}$$

$$\frac{\partial^2\psi}{\partial x^2} = (\frac{\partial\psi'_r}{\partial r}\frac{\partial r}{\partial x} + \frac{\partial\psi'_r}{\partial\theta}\frac{\partial\theta}{\partial x} + \frac{\partial\psi'_r}{\partial\phi}\frac{\partial\phi}{\partial x})\frac{\partial r}{\partial x} +$$

$$(\frac{\partial\psi'_\theta}{\partial r}\frac{\partial r}{\partial x} + \frac{\partial\psi'_\theta}{\partial\theta}\frac{\partial\theta}{\partial x} + \frac{\partial\psi'_\theta}{\partial\phi}\frac{\partial\phi}{\partial x})\frac{\partial\theta}{\partial x} +$$

$$(\frac{\partial\psi'_\phi}{\partial r}\frac{\partial r}{\partial x} + \frac{\partial\psi'_\phi}{\partial\theta}\frac{\partial\theta}{\partial x} + \frac{\partial\psi'_\phi}{\partial\phi}\frac{\partial\phi}{\partial x})\frac{\partial\phi}{\partial x}$$

$$\frac{\partial^2\psi}{\partial x^2} = \frac{\partial^2\psi}{\partial r^2}(\frac{\partial r}{\partial x})^2 + \frac{\partial^2\psi}{\partial\theta\partial r}\frac{\partial\theta}{\partial x}\frac{\partial r}{\partial x} + \frac{\partial^2\psi}{\partial\phi\partial r}\frac{\partial\phi}{\partial x}\frac{\partial r}{\partial x} +$$

$$(\frac{\partial^2\psi}{\partial r\partial\theta}\frac{\partial r}{\partial x}\frac{\partial\theta}{\partial x} + \frac{\partial^2\psi}{\partial\theta^2}(\frac{\partial\theta}{\partial x})^2 + \frac{\partial^2\psi}{\partial\phi\partial\theta}\frac{\partial\phi}{\partial x}\frac{\partial\theta}{\partial x} +$$

$$(\frac{\partial^2\psi}{\partial r\partial\phi}\frac{\partial r}{\partial x}\frac{\partial\phi}{\partial x} + \frac{\partial^2\psi}{\partial\theta\partial\phi}\frac{\partial\theta}{\partial x}\frac{\partial\phi}{\partial x} + \frac{\partial^2\psi}{\partial\phi^2}(\frac{\partial\phi}{\partial x})^2$$

Similarly $\frac{\partial^2\psi}{\partial y^2}$ and $\frac{\partial^2\psi}{\partial z^2}$ can be written by symmetry.

Then the corresponding values in Polar coordinate format be substituted for those in the Cartesian format as was done in the case of calculation of angular momentum earlier.

8.11 Supplement to Rigid Rotor

$$z = \cos\theta, \quad \frac{dz}{d\theta} = -\sin\theta, \quad \left(\frac{dz}{d\theta}\right)^2 = \sin^2\theta$$

$$\frac{d^2z}{d\theta^2} = -\cos\theta, \quad \frac{d}{d\theta} = -\sin\theta\frac{d}{dz}$$

$$\frac{d^2}{d\theta^2} = \frac{d}{d\theta}\left(\frac{d}{d\theta}\right) = \frac{d}{d\theta}\left(\frac{dz}{d\theta}\frac{d}{dz}\right)$$

$$= \frac{d}{dz}\frac{d^2z}{d\theta^2} + \frac{dz}{d\theta}\left(\frac{d}{d\theta}\frac{d}{dz}\right) = \frac{d}{dz}\frac{d^2z}{d\theta^2} + \frac{dz}{d\theta}\left(\frac{dz}{d\theta}\frac{d}{dz}\right)\frac{d}{dz}$$

$$\frac{d^2}{d\theta^2} = -\cos\theta\frac{d}{dz} + \sin^2\theta\frac{d^2}{dz^2}$$

$$\therefore \frac{d^2y}{d\theta^2} = -\cos\theta\frac{dy}{dz} + \sin^2\theta\frac{d^2y}{dz^2}$$

$$\frac{dy}{d\theta} = -\sin\theta\frac{dy}{dz}$$

Associated Legendre function

$$(1-z^2)\frac{d^2 P_l^m(z)}{dz^2} - 2z\frac{dp_l^m(z)}{dz} + \left[l(l+1) - \frac{m^2}{1-z^2}\right]P_l^m(z) = 0$$

Associated Legendre polynomial

$$P_l^{-m}(z) = (-1)^m \frac{(l-m)!}{(l+m)!} P_l^m(z) \quad \text{where } 1 = 0, 1, 2, 3...$$

where $m \geq 0$ and $|m| \leq l$

8.12 Supplement to One-dimensional Harmonic Oscillator

$$y = x\sqrt{\beta}, \quad \frac{dy}{dx} = \sqrt{\beta}, \quad \frac{d}{dx} = \frac{d}{dy}\sqrt{\beta}, \quad \frac{d^2}{dx^2} = \frac{d^2}{dy^2}\beta$$

$$\therefore \beta\frac{d^2\psi}{dy^2} + \left(\alpha - \beta^2\frac{y^2}{\beta}\right)\psi = 0$$

152   Quantum Chemistry

$$\frac{d^2\psi}{dy^2} + \left(\frac{\alpha}{\beta} - y^2\right)\psi = 0$$

$$y = x\sqrt{\beta}, \quad \frac{dy}{dx} = \sqrt{\beta}, \quad \frac{d}{dx} = \frac{d}{dy}\sqrt{\beta}, \quad \frac{d^2}{dx^2} = \frac{d^2}{dy^2}\beta$$

$$\therefore \beta\frac{d^2\psi}{dy^2} + \left(\alpha - \beta^2\frac{y^2}{\beta}\right)\psi = 0$$

$$\frac{d^2\psi}{dy^2} + \left(\frac{\alpha}{\beta} - y^2\right)\psi = 0$$

$$\psi = f \exp\left(\tfrac{-1}{2}y^2\right)$$

$$\therefore \frac{d\psi}{dy} = f(-y)\exp\left(\tfrac{-1}{2}y^2\right) + \exp\left(\tfrac{-1}{2}y^2\right)\frac{df}{dy}$$

$$\frac{d^2\psi}{dy^2} = f(y^2)\exp\left(\tfrac{-1}{2}y^2\right) + \exp\left(\tfrac{-1}{2}y^2\right)\left\{-y\frac{df}{dy} - f\right\}$$

$$+ \frac{d^2f}{dy^2}\exp\left(\tfrac{-1}{2}y^2\right) - y\frac{df}{dy}\exp\left(\tfrac{-1}{2}y^2\right)$$

$$= \exp\left(\tfrac{-1}{2}y^2\right)\left[\frac{d^2f}{dy^2} - 2y\frac{df}{dy} + y^2 f - f\right]$$

$$\therefore \frac{d^2\psi}{dy^2} + \left(\frac{\alpha}{\beta} - y^2\right)\psi = \left(\frac{d^2f}{dy^2} - 2y\frac{df}{dy} + y^2 f - f\right)\exp\left(\tfrac{-1}{2}y^2\right)$$

$$+ \left(\frac{\alpha}{\beta} - y^2\right)f\exp\left(\tfrac{-1}{2}y^2\right) = 0$$

$$= \left[\frac{d^2f}{dy^2} - 2y\frac{df}{dy} + \left(\frac{\alpha}{\beta} - 1\right)f\right]\exp\left(\tfrac{-1}{2}y^2\right) = 0$$

$$\therefore \frac{d^2f}{dy^2} - 2y\frac{df}{dy} + \left(\frac{\alpha}{\beta} - 1\right)f = 0$$

Hermite's equation

$$\frac{d^2}{dy^2}H_n(y) - 2y\frac{d}{dy}H_n(y) + 2H_n(y) = 0$$